Fam

A report o ... rvey

Editor: Anthony Craggs

London: TSO

ISBN 0 11 621734 0
ISSN 0965-1403

Contact points
For enquiries about this publication, contact the Editor:
Anthony Craggs

Tel: **020 7533 5760**
E-mail: **anthony.craggs@ons.gsi.gov.uk**

To order this publication, contact TSO on 0870 600 5522.
See also back cover.

For general enquiries, contact the National Statistics Customer Contact Centre on **0845 601 3034**
(minicom: 01633 812399)
E-mail: info@statistics.gsi.gov.uk
Fax: 01633 652747
Letters: Room D115, Government Buildings,
Cardiff Road, Newport NP10 8XG

You can also find National Statistics on the Internet at
www.statistics.gov.uk

About the Office for National Statistics
The Office for National Statistics (ONS) is the government agency responsible for compiling, analysing and disseminating many of the United Kingdom's economic, social and demographic statistics, including the retail prices index, trade figures and labour market data, as well as the periodic census of the population and health statistics. The Director of ONS is also the National Statistician and the Registrar General for England and Wales, and the agency administers the registration of births, marriages and deaths there.

A National Statistics Publication
National Statistics are produced to high professional standards set out in the National Statistics Code of Practice. They undergo regular quality assurance reviews to ensure that they meet customer needs. They are produced free from any political interference.

Contents

			Page
Introduction			**5**
1	**Expenditure by income**		**9**
Tables	1.1	main items by gross income decile	14
	1.2	percentage on main items by gross income decile	16
	1.3	detailed expenditure by gross income decile	18
	1.4	main items by disposable income decile	30
	1.5	percentage on main items by disposable income decile	32
2	**Expenditure by age and income**		**35**
Tables	2.1	main items for all age groups	40
	2.2	main items as a percentage for all age groups	41
	2.3	detailed expenditure for all age groups	42
	2.4	aged under 30 by income	48
	2.5	aged 30 and under 50 by income	49
	2.6	aged 50 and under 65 by income	50
	2.7	aged 65 and under 75 by income	51
	2.8	aged 75 or over by income	52
3	**Expenditure by socio-economic characteristics**		**53**
Tables	3.1	by economic activity status	56
	3.2	full-time employee by income	58
	3.3	self-employed by income	59
	3.4	by number of persons working	60
	3.5	by age completed continuous full-time education	61
	3.6	by socio-economic class	62
4	**Expenditure by household composition, income and tenure**		**65**
Tables	4.1	by household composition	70
	4.2	one person retired households mainly dependent on state pensions	72
	4.3	one person retired households not mainly dependent on state pensions	73
	4.4	one person non-retired	74
	4.5	one adult with children	75
	4.6	two adults with children	76
	4.7	one man one woman non-retired	77
	4.8	one man one woman retired mainly dependent on state pensions	78
	4.9	one man one woman retired not mainly dependent on state pensions	79
	4.10	by tenure	80
	Map of regions		82
5	**Expenditure by region**		**83**
Tables	5.1	main items of expenditure	88
	5.2	main items as a percentage of expenditure	90
	5.3	detailed expenditure	92
	5.4	expenditure by urban/rural areas (GB only)	104

6 **Trends in household expenditure** **105**

Tables 6.1 main items 1978 - 2002-03 ---------- 106
6.2 percentage on main items 1978 - 2002-03 ---------- 108

7 **Detailed expenditure, recent changes & place of purchase** **113**

Tables 7.1 with full method standard errors ---------- 118
7.2 alcoholic drink by type of premises ---------- 128
7.3 food by place of purchase ---------- 130
7.4 selected items by place of purchase ---------- 132
7.5 clothing and footwear by place of purchase ---------- 133

8 **Household income** **135**

Tables 8.1 by household composition ---------- 140
8.2 by age of household reference person ---------- 140
8.3 by income group ---------- 141
8.4 by household tenure ---------- 141
8.5 by region ---------- 142
8.6 by GB urban/rural areas ---------- 142
8.7 by socio-economic class ---------- 143
8.8 1970 to 2002-03 ---------- 143

9 **Household characteristics and ownership of durable goods** **145**

Tables 9.1 households ---------- 150
9.2 persons ---------- 152
9.3 durable goods 1970 to 2002-03 ---------- 152
9.4 durable goods by income & household composition ---------- 153
9.5 with cars ---------- 154
9.6 durable goods by region ---------- 155
9.7 size, composition & age by income group ---------- 156
9.8 economic activity, tenure & socio-economic class ---------- 158

Appendices **161**

A Description and response rate of the survey ---------- 162
B Uses of the survey ---------- 166
C Standard errors and estimates of precision ---------- 168
D Definitions ---------- 173
E Changes in definitions 1991 to 2002-03 ---------- 186
F Differential grossing ---------- 191
G Index to tables in reports on the FES/EFS in 1994-95 to 2002-03 ---------- 193

Family Spending

The Expenditure and Food Survey

The 2002-03 survey

Data quality and definitions

Related data sources

Additional tabulations

Acknowledgements

Contact points

Symbols and conventions used in this report

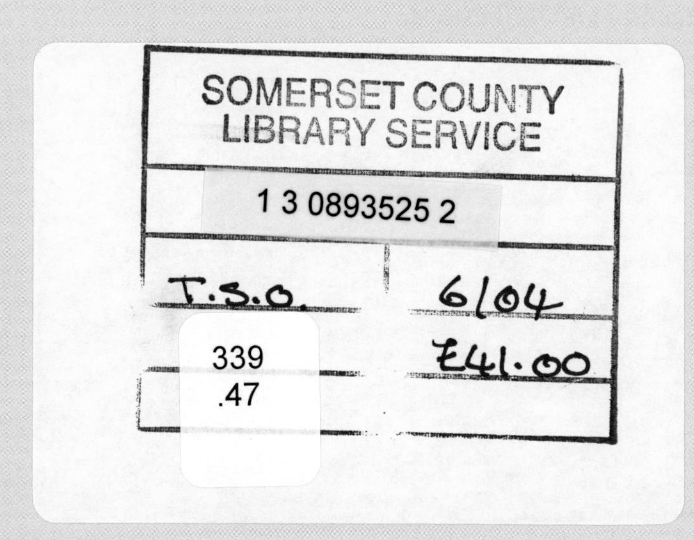

Introduction

The Expenditure and Food Survey

This report presents the latest information from the Expenditure and Food Survey (EFS) for the financial year April 2002 to March 2003.

The EFS is the result of the amalgamation of the Family Expenditure and National Food Surveys (FES and NFS). Both surveys were well established and important sources of information for government and the wider community, charting changes and patterns in Britain's spending and food consumption since the 1950s. The Office for National Statistics (ONS) has overall project management and financial responsibility for the EFS while the Department for Environment, Food and Rural Affairs (DEFRA) sponsors the specialist food data.

The design of the EFS is based on the FES and the same questions were asked of the respondents. The survey continues to be primarily used to provide information for the Retail Prices Index; National Accounts estimates of household expenditure; the analysis of the effect of taxes and benefits, and trends in nutrition. However, the results are multi purpose, providing an invaluable supply of economic and social data.

The 2002-03 survey

In 2002-03 6,342 households in Great Britain took part in the EFS. The response rate was 58 per cent in Great Britain and 56 per cent in Northern Ireland. The fieldwork was undertaken by the Office for Ntional Statistics and the Northern Ireland Statistics and Research Agency.

Further details about the conduct of the survey are given in Appendix A.

Data quality and definitions

The results shown in this report are of the data collected by the EFS, following a process of validation and adjustment for non-response using weights that control for a number of factors. These issues are discussed in the section on reliability in Appendix A.

Figures in the report are subject to sampling variability. Standard errors for detailed expenditure items are presented in relative terms in **table 7.1** and are described in **Appendix C**. Figures shown for particular groups of households (e.g. income groups or household composition groups), regions or other sub-sets of the sample are subject to larger sampling variability, and are more sensitive to possible extreme values than are figures for the sample as a whole. In order to increase reliability, some tables are based on data collected across two years.

The definitions used in the report are set out in Appendix D, and changes made since 1991 are described in Appendix E. Note particularly that housing benefit and council tax rebate (rates rebate in Northern Ireland), unlike other social security benefits, are not included in income but are shown as a reduction in housing costs.

Related data sources

Details of household consumption expenditure within the context of the UK National Accounts are produced as part of *Consumer Trends* (http://www.statistics.gov.uk/downloads/theme_economy/Consumer_Trends_Q3_2003.pdf). It includes all expenditure by members of UK resident households. National Accounts figures draw on a number of sources including the EFS: figures shown in this report are therefore not directly comparable to National Accounts data. National Accounts data may be more appropriate for deriving long term trends on expenditure.

More detailed income information is available from the Family Resources Survey (FRS), conducted for the Department for Work and Pensions. Further information about food consumption, and in particular details of food quantities, is available from the Department for Environment, Food and Rural Affairs, who are continuing to produce their own report of the survey.

In Northern Ireland, a companion survey to the GB EFS is conducted by the Central Survey Unit of the Northern Ireland Statistics and Research Agency (NISRA). Households in Northern Ireland are over-sampled so that separate analysis can be carried out, however these cases are downweighted when UK data are analysed. Results from this sample will be published in a separate report, *The Northern Ireland Family Expenditure Survey Report for 2002-03.* Further information and copies of this report can be obtained from:

Northern Ireland Statistics and Research Agency,
Central Survey Unit
McAuley House
2-14 Castle Street
Belfast BT1 1SY

Tel: 02890 348 215

Additional tabulations

The report gives a broad overview of the results of the survey, and provides more detailed information about some aspects of expenditure. However, many users of EFS data have very specific data requirements that may not appear in the desired form in this report. The ONS can provide more detailed analysis of the tables in this report, and can also provide additional tabulations to meet specific requests. A charge will be made to cover the cost of providing additional information.

The tables in Family Spending 2002-03 are available as Excel spreadsheets (with unrounded data).

Acknowledgements

A large scale survey is a collaborative effort and the authors wish to thank the interviewers and other ONS staff who contributed to the study. The survey would not be possible without the co-operation of the respondents who gave up their time to be interviewed and keep a diary of their spending. Their help is gratefully acknowledged.

Contact points

Please address all enquiries to:

Expenditure and Food Survey,
Office for National Statistics,
Room D1/23,
1, Drummond Gate,
London SW1V 2QQ.

Tel: 020 7533 5756 (answering machine outside office hours)
Fax: 020 7533 5300

Symbols and conventions used in this report

.. Data not available due to unreliability, as a result of:

1. too few reporting households, generally less than 10, or
2. sampling error too large, generally 50 per cent or more

[ ] Figures to be used with extra caution because based on fewer than 20 reporting households.

Rounding: Individual figures have been rounded independently. The sum of component items does not therefore necessarily add to the totals shown.

Averages: These are averages (means) for all households included in the column or row, and are not restricted to those households reporting expenditure on a particular item or income of a particular type.

Period covered: Financial year 2002-03 (1 April 2002 to 31 March 2003).

Chapter 1

Expenditure by income

Income is not adjusted to take into account the different composition of households (equivalisation) as done in some other income analyses

- Average weekly expenditure in 2002-03 was £406. It ranged from £136 a week in the lowest of the ten income groups to £883 a week in the highest.

- **Transport** was the highest category of spending, with an average of £59 a week. Next was **recreation and culture**, £56 a week, followed by **food and non-alcoholic drinks**, £43 a week.

- For households in the lowest income range the highest categories of spending were **food and non-alcoholic drinks** and **housing, fuel and power** (excluding mortgage interest payments, council tax/rates). For households in the upper five deciles, however, the highest expenditure was for **transport** and **recreation and culture.**

- As a proportion of their total expenditure, households in the lowest income group spent twice as much (16 per cent) on **food and non-alcoholic drinks** as those in the highest income group (8 per cent).

- Spending on **tobacco** was highest for households in the middle income groups who spent up to £7.00 a week compared to £3.80 for those in the lowest group and £5.10 in the highest group.

- The proportion of **transport** spending going on the purchase of vehicles increased with income, from around 30 per cent in the lower income groups to nearly 50 per cent in the higher groups.

- The proportion of **recreation and culture** spending going on **gambling** was higher among low income groups (10 per cent) than higher income groups (four per cent) although the latter spent more in absolute terms.

1.1 Average weekly expenditure on the main commodities and services

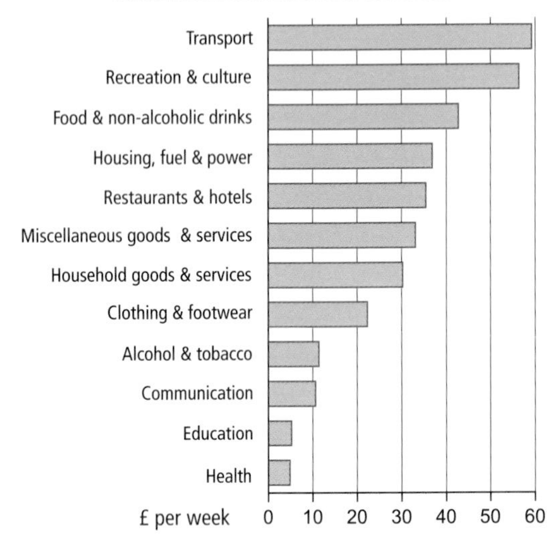

1.2 Average weekly expenditure on the main commodities and services

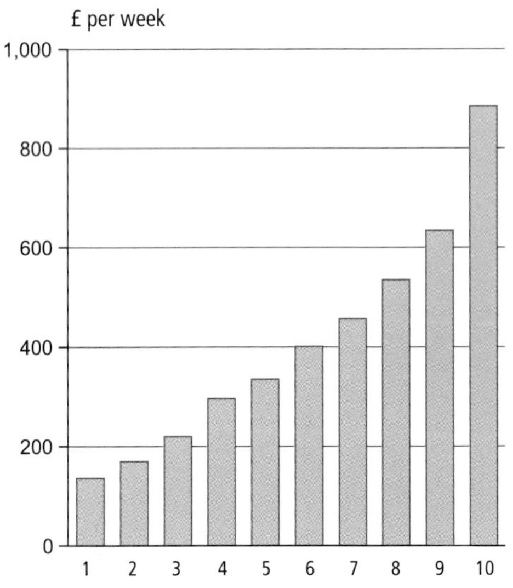

Chapter 1
Expenditure by income

Tables 1.1 to 1.3 show household expenditure on commodities and services by gross income group. The figures include information from the expenditure diaries kept by adults and children aged 7 to 15. The expenditure categories in this year's report are comparable with the 2001-02 report, but are not directly comparable with previous years. This is due to the adoption of the European standard Classification of Individual Consumption by Purpose (COICOP) in 2001.

Ten income groups are shown, based on gross income, with an equal number of households in each group (decile). The characteristics of households varied across income groups. Differences in spending may therefore be the result of other factors as well as income. Household size is particularly important in determining spending levels. The average number of persons per household is provided for each income group. The figures show that as income increases, the average number of people in each household also increases. The highest income group had more than twice the average household size of the lowest income group (3.2 people compared with 1.3 people).

Tables 1.4 and **1.5** show how expenditure varies with disposable income instead of gross income. That is, gross income *less* income tax and National Insurance contributions.

Total household expenditure

In 2002-03 total expenditure was made up from the total of COICOP expenditure groups (codes 1-12) plus other expenditure items (code 13). Other expenditure items were all the items included in the context of *Family Spending* expenditure but excluded from COICOP. For a breakdown of these items see **Table 7.1** on page 118.

Figure 1.1 shows the overall distribution of expenditure between the main commodities and services. Transport was the largest item of spending in 2002-03 at £59 a week, followed by recreation and culture at £56. In general, expenditure across all commodities and services increased with income.

Table 1.1 and Figure 1.2 show that the average weekly expenditure ranged from £136 per week in the lowest income decile group to £883 per week in the highest. The highest income group spent more than double the average expenditure for all UK households.

Expenditure patterns

Transport was the largest area of expenditure, at £59 a week, closely followed by recreation and culture at £56 a week, and food and non-alcoholic drink at £43 a week. **Figures 1.3a** and **1.3b** show average weekly expenditure by gross income decile group for selected commodities and services. There was considerable variation between income groups in the level of expenditure. For example, expenditure on transport ranged from £13 a week for the lowest income group, to £140 a week for the highest, spending on food ranged from £21 a week to £67 a week in the highest group, while spending on restaurants and hotels ranged from £10 to £83.

Table 1.2 shows expenditure on each commodity as a percentage of total expenditure. Spending on some commodities increased broadly in line with total expenditure, so that the proportion spent on them did not vary much with income. However, it shows that the proportion spent on education was steady for the lowest nine income decile groups, but increased threefold for the highest income group to three per cent. **Figure 1.4** shows that the proportion spent on transport increased with income ranging from nine per cent for the second lowest income group to 17 per cent for the second highest. In contrast, the proportion spent on food and non-alcoholic drinks decreased with income, ranging from 16 per cent for lower incomes, to eight per cent for the highest. The proportion spent on housing, fuel and power also broadly decreased, from 17 per cent in the second income group, to seven per cent in the highest.

1.3a Food, housing and restaurants by gross income decile group

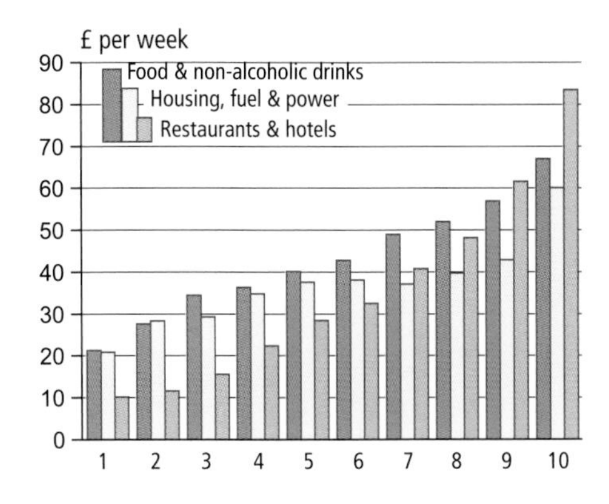

1.3b Transport and recreation & culture by gross income decile group

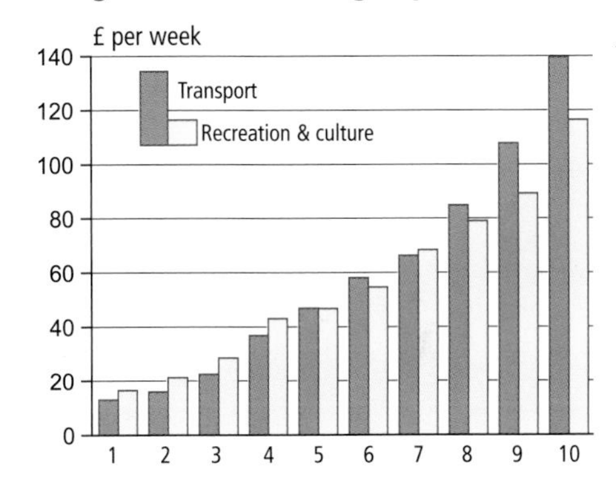

1.4 Food, housing and transport as a percentage of total expenditure by gross income decile group

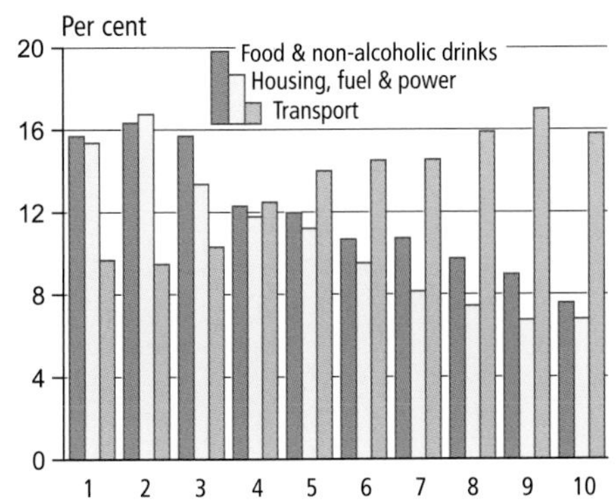

1.5 Tobacco, wine and beer by gross income decile group

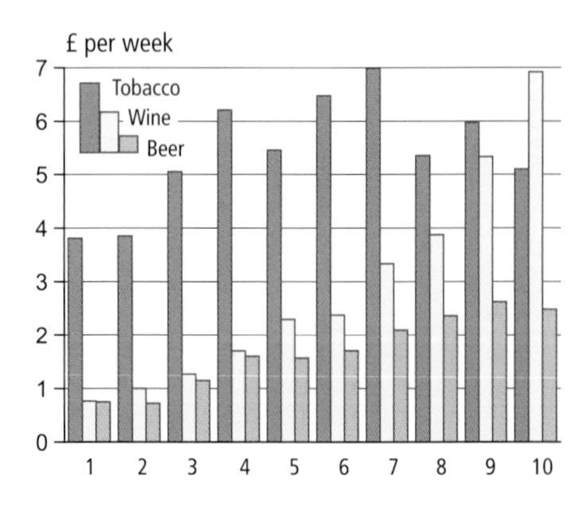

1.6 Restaurant and takeaway meals by gross income decile group

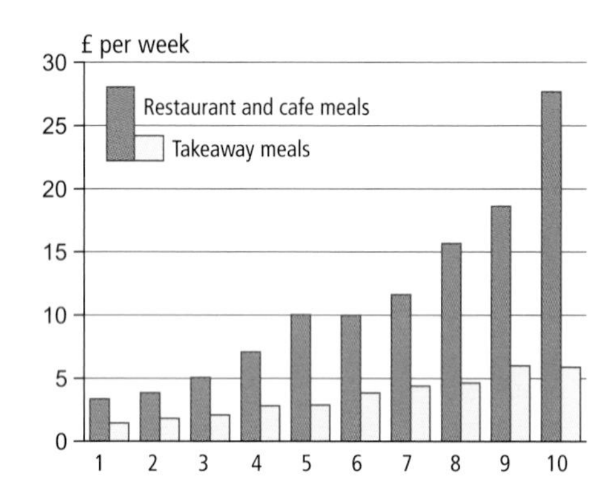

1.7 Rail and bus fares as a percentage of expenditure on transport services by gross income decile group

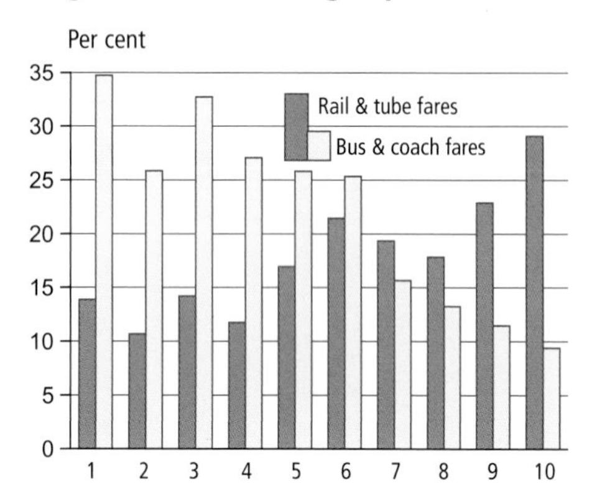

Detailed expenditure patterns

Table 1.3 shows a more detailed breakdown of expenditure categories for the income groups.

Alcoholic drink and tobacco

Average weekly expenditure on tobacco did not increase with income unlike most other items. The highest expenditure was by the seventh decile group which spent £7.00 a week compared with the lowest and highest groups at £3.80 and £5.10 respectively. Total spending on alcoholic drink brought home increased with income. However, **Figure 1.5** shows that expenditure on selected drinks varied. The lowest income group spent a similar amount on wine and beer, at 80p and 70p a week respectively. In contrast, the highest income group spent £6.90 a week on wine and £2.50 on beer.

Catering services

Figure 1.6 compares the expenditure on restaurant and café meals, with that on takeaway meals. While expenditure on both categories increased with income, the difference between income groups are much starker for expenditure on restaurants and café meals, where households in the highest income group spent over eight times that spent by those in the lowest income group.

Fares and other travel costs

Figure 1.7 highlights variations in expenditure on transport services, which includes rail, tube, bus and coach fares, taxis, air and combined tickets. As a percentage of this total, expenditure on rail and tube fares generally increased with income, with the highest income group spending 29 per cent on rail and tube fares. However the proportion spent on bus and coach fares generally decreased with income, with the lowest income group spending 35 per cent of fares money on bus and coach fares and the highest income group spending only nine per cent, although they spent more in absolute terms.

Recreational and cultural services

Figure 1.8 shows expenditure on sports admissions, cinemas and the provision of television-related services (TV, video, satellite rentals, cable subscriptions, TV licences and internet services) as a percentage of total expenditure on recreational and cultural services. As a proportion of expenditure on recreation services, households in the lowest income groups spent almost 40 per cent on the provision of television-related services, compared to under 20 per cent for the highest income group. The proportion of spending on sports activities generally increased with income, ranging from 12 to 49 per cent. Expenditure on cinemas, theatre and museums also increased, although to a lesser extent, ranging from 5 to 13 per cent of the total on recreational and cultural services.

1.8 Sports, cinemas and electronic services as a percentage of expenditure on recreational services by gross income decile group

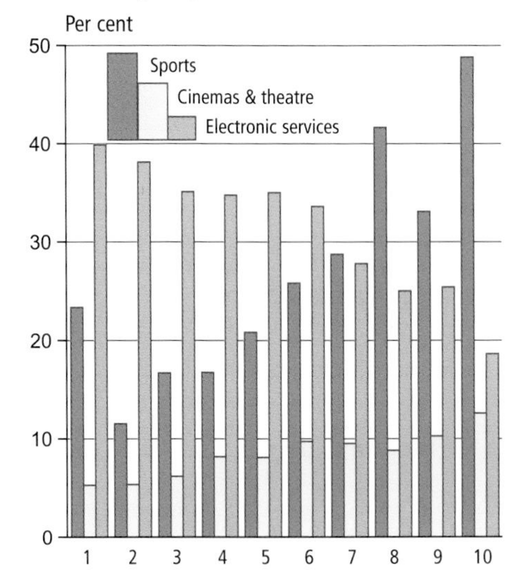

Expenditure by disposable income

Tables 1.4 and **1.5** show how expenditure varied with disposable income, that is, gross income less income tax and National Insurance contributions. Some households will be in a different income decile when defined by disposable income rather than gross income. If this is the case, they will normally move only one group up or down. **Figure 1.9** shows that the variation of total expenditure with income depended very little on the measure of income used. Comparisons of **Table 1.1** with **1.4** and of **1.2** with **1.5** show that the pattern of expenditure was also similar.

1.9 Average weekly expenditure by gross and disposable income decile group

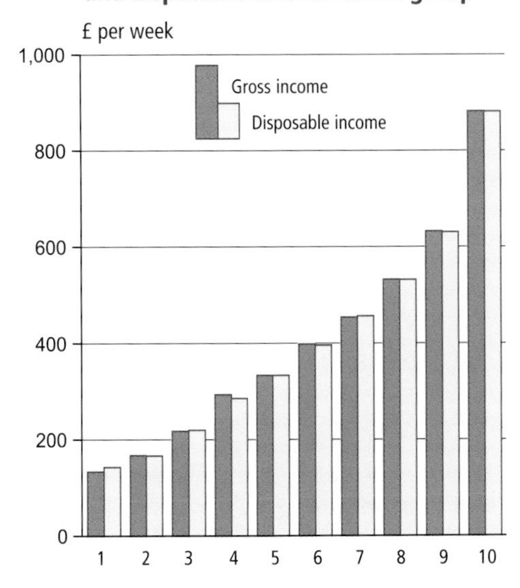

1.1 Household expenditure by gross income decile group 2002-03

based on weighted data and including children's expenditure

		Lowest ten per cent	Second decile group	Third decile group	Fourth decile group	Fifth decile group	Sixth decile group
Lower boundary of group (£ per week)			123	188	259	341	435
Grossed number of households (thousands)		2,440	2,430	2,440	2,430	2,440	2,440
Total number of households in sample		701	724	732	715	719	697
Total number of persons in sample		901	1,278	1,412	1,590	1,718	1,807
Total number of adults in sample		774	967	1,155	1,189	1,282	1,295
Weighted average number of persons per household		1.3	1.7	1.9	2.2	2.4	2.5
Commodity or service		**Average weekly household expenditure (£)**					
1	**Food & non-alcoholic drinks**	21.30	27.70	34.50	36.30	40.10	42.80
2	**Alcoholic drinks, tobacco & narcotics**	5.90	6.30	8.80	10.50	10.60	12.00
3	**Clothing & footwear**	6.20	7.90	11.40	15.80	18.00	20.20
4	**Housing[1], fuel & power**	20.90	28.40	29.30	34.80	37.60	38.10
5	**Household goods & services**	11.00	12.70	17.60	23.20	22.00	29.80
6	**Health**	1.60	2.20	4.40	3.10	4.50	5.50
7	**Transport**	13.10	16.10	22.60	36.90	46.90	58.10
8	**Communication**	5.40	5.90	7.50	8.80	9.50	10.70
9	**Recreation & culture**	16.60	21.30	28.60	43.10	46.70	54.70
10	**Education**	[0.70]	1.70	0.70	1.20	2.00	3.60
11	**Restaurants & hotels**	10.20	11.60	15.50	22.30	28.50	32.50
12	**Miscellaneous goods & services**	10.20	12.70	16.00	22.10	25.90	33.80
1-12	**All expenditure groups**	123.10	154.50	196.90	258.30	292.30	341.80
13	**Other expenditure items**	12.50	14.90	22.50	36.90	42.70	58.60
Total expenditure		135.60	169.40	219.40	295.20	335.00	400.40
Average weekly expenditure per person (£)							
Total expenditure		106.00	98.40	114.40	134.80	140.80	157.40

Note: The commodity and service categories are not comparable with those in publications before 2001-02
1 Excluding mortgage interest payments, council tax and Northern Ireland rates

1.1 Household expenditure by gross income decile group (cont.) 2002-03

based on weighted data and including children's expenditure

	Seventh decile group	Eighth decile group	Ninth decile group	Highest ten per cent	All house-holds
Lower boundary of group (£ per week)	541	662	821	1,085	
Grossed number of households (thousands)	2,430	2,440	2,430	2,430	24,350
Total number of households in sample	687	672	655	625	6,927
Total number of persons in sample	1,967	1,917	1,974	2,022	16,586
Total number of adults in sample	1,409	1,407	1,477	1,495	12,450
Weighted average number of persons per household	2.8	2.8	3.0	3.2	2.4
Commodity or service	**Average weekly household expenditure (£)**				
1 Food & non-alcoholic drinks	**49.00**	**52.00**	**56.90**	**66.90**	**42.70**
2 Alcoholic drinks, tobacco & narcotics	**14.10**	**13.10**	**15.80**	**16.40**	**11.40**
3 Clothing & footwear	**26.20**	**31.00**	**36.60**	**49.60**	**22.30**
4 Housing[1], fuel & power	**37.10**	**39.70**	**42.80**	**60.00**	**36.90**
5 Household goods & services	**33.00**	**36.70**	**44.40**	**71.90**	**30.20**
6 Health	**5.20**	**6.50**	**5.30**	**9.90**	**4.80**
7 Transport	**66.30**	**84.80**	**107.80**	**139.60**	**59.20**
8 Communication	**11.30**	**13.60**	**15.00**	**18.30**	**10.60**
9 Recreation & culture	**68.40**	**79.00**	**89.30**	**116.40**	**56.40**
10 Education	**3.40**	**4.30**	**8.50**	**25.80**	**5.20**
11 Restaurants & hotels	**40.80**	**48.10**	**61.50**	**83.40**	**35.40**
12 Miscellaneous goods & services	**37.40**	**45.10**	**51.20**	**76.50**	**33.10**
1-12 All expenditure groups	**392.30**	**454.10**	**535.20**	**734.90**	**348.30**
13 Other expenditure items	**63.60**	**80.00**	**99.00**	**148.50**	**57.90**
Total expenditure	**455.90**	**534.10**	**634.20**	**883.40**	**406.20**
Average weekly expenditure per person (£)					
Total expenditure	**161.60**	**192.20**	**213.60**	**274.40**	**170.50**

Note: The commodity and service categories are not comparable with those in publications before 2001-02
1 Excluding mortgage interest payments, council tax and Northern Ireland rates

1.2 Household expenditure as a percentage of total expenditure by gross income decile group

2002-03

based on weighted data and including children's expenditure

		Lowest ten per cent	Second decile group	Third decile group	Fourth decile group	Fifth decile group	Sixth decile group
Lower boundary of group (£ per week)			123	188	259	341	435
Grossed number of households (thousands)		2,440	2,430	2,440	2,430	2,440	2,440
Total number of households in sample		701	724	732	715	719	697
Total number of persons in sample		901	1,278	1,412	1,590	1,718	1,807
Total number of adults in sample		774	967	1,155	1,189	1,282	1,295
Weighted average number of persons per household		1.3	1.7	1.9	2.2	2.4	2.5
Commodity or service		**Percentage of total expenditure**					
1	**Food & non-alcoholic drinks**	*16*	*16*	*16*	*12*	*12*	*11*
2	**Alcoholic drinks, tobacco & narcotics**	*4*	*4*	*4*	*4*	*3*	*3*
3	**Clothing & footwear**	*5*	*5*	*5*	*5*	*5*	*5*
4	**Housing[1], fuel & power**	*15*	*17*	*13*	*12*	*11*	*10*
5	**Household goods & services**	*8*	*7*	*8*	*8*	*7*	*7*
6	**Health**	*1*	*1*	*2*	*1*	*1*	*1*
7	**Transport**	*10*	*9*	*10*	*12*	*14*	*15*
8	**Communication**	*4*	*3*	*3*	*3*	*3*	*3*
9	**Recreation & culture**	*12*	*13*	*13*	*15*	*14*	*14*
10	**Education**	*[1]*	*1*	*0*	*0*	*1*	*1*
11	**Restaurants & hotels**	*8*	*7*	*7*	*8*	*8*	*8*
12	**Miscellaneous goods & services**	*8*	*8*	*7*	*7*	*8*	*8*
1-12	**All expenditure groups**	*91*	*91*	*90*	*88*	*87*	*85*
13	**Other expenditure items**	*9*	*9*	*10*	*12*	*13*	*15*
Total expenditure		*100*	*100*	*100*	*100*	*100*	*100*

Note: The commodity and service categories are not comparable with those in publications before 2001-02
1 Excluding mortgage interest payments, council tax and Northern Ireland rates

1.2 Household expenditure as a percentage of total expenditure by gross income decile group (cont.) 2002-03

based on weighted data and including children's expenditure

	Seventh decile group	Eighth decile group	Ninth decile group	Highest ten per cent	All house-holds
Lower boundary of group (£ per week)	541	662	821	1085	
Grossed number of households (thousands)	2,430	2,440	2,430	2,430	24,350
Total number of households in sample	687	672	655	625	6,927
Total number of persons in sample	1,967	1,917	1,974	2,022	16,586
Total number of adults in sample	1,409	1,407	1,477	1,495	12,450
Weighted average number of persons per household	2.8	2.8	3.0	3.2	2.4
Commodity or service		**Percentage of total expenditure**			
1 **Food & non-alcoholic drinks**	*11*	*10*	*9*	*8*	*11*
2 **Alcoholic drinks, tobacco & narcotics**	*3*	*2*	*2*	*2*	*3*
3 **Clothing & footwear**	*6*	*6*	*6*	*6*	*5*
4 **Housing[1], fuel & power**	*8*	*7*	*7*	*7*	*9*
5 **Household goods & services**	*7*	*7*	*7*	*8*	*7*
6 **Health**	*1*	*1*	*1*	*1*	*1*
7 **Transport**	*15*	*16*	*17*	*16*	*15*
8 **Communication**	*2*	*3*	*2*	*2*	*3*
9 **Recreation & culture**	*15*	*15*	*14*	*13*	*14*
10 **Education**	*1*	*1*	*1*	*3*	*1*
11 **Restaurants & hotels**	*9*	*9*	*10*	*9*	*9*
12 **Miscellaneous goods & services**	*8*	*8*	*8*	*9*	*8*
1-12 **All expenditure groups**	*86*	*85*	*84*	*83*	*86*
13 **Other expenditure items**	*14*	*15*	*16*	*17*	*14*
Total expenditure	*100*	*100*	*100*	*100*	*100*

Note: The commodity and service categories are not comparable with those in publications before 2001-02
1 Excluding mortgage interest payments, council tax and Northern Ireland rates

1.3 Detailed household expenditure by gross income decile group 2002-03

based on weighted data and including children's expenditure

		Lowest ten per cent	Second decile group	Third decile group	Fourth decile group	Fifth decile group	Sixth decile group
Lower boundary of group (£ per week)			123	188	259	341	435
Grossed number of households (thousands)		2,440	2,430	2,440	2,430	2,440	2,440
Total number of households in sample		701	724	732	715	719	697
Total number of persons in sample		901	1,278	1,412	1,590	1,718	1,807
Total number of adults in sample		774	967	1,155	1,189	1,282	1,295
Weighted average number of persons per household		1.3	1.7	1.9	2.2	2.4	2.5
Commodity or service		**Average weekly household expenditure (£)**					
1	**Food & non-alcoholic drinks**	**21.30**	**27.70**	**34.50**	**36.30**	**40.10**	**42.80**
1.1	Food	19.50	25.50	31.60	33.20	36.60	39.20
1.1.1	Bread, rice and cereals	1.90	2.60	3.00	3.40	3.70	4.00
1.1.2	Pasta products	0.20	0.20	0.20	0.30	0.30	0.30
1.1.3	Buns, cakes, biscuits etc.	1.40	1.90	2.50	2.50	2.80	2.80
1.1.4	Pastry (savoury)	0.20	0.30	0.30	0.40	0.50	0.70
1.1.5	Beef (fresh, chilled or frozen)	0.60	0.90	1.20	1.10	1.20	1.50
1.1.6	Pork (fresh, chilled or frozen)	0.30	0.40	0.50	0.60	0.60	0.60
1.1.7	Lamb (fresh, chilled or frozen)	0.30	0.40	0.50	0.50	0.50	0.60
1.1.8	Poultry (fresh, chilled or frozen)	0.70	0.80	1.10	1.20	1.40	1.40
1.1.9	Bacon and ham	0.50	0.60	0.80	0.80	0.90	0.90
1.1.10	Other meat and meat preparations	2.50	3.20	3.80	4.20	4.40	5.00
1.1.11	Fish and fish products	1.00	1.20	1.60	1.50	1.70	1.50
1.1.12	Milk	1.40	1.70	1.90	2.00	2.10	2.30
1.1.13	Cheese and curd	0.60	0.80	1.00	1.00	1.20	1.40
1.1.14	Eggs	0.30	0.30	0.40	0.40	0.40	0.40
1.1.15	Other milk products	0.70	0.90	1.00	1.10	1.30	1.40
1.1.16	Butter	0.20	0.20	0.30	0.20	0.30	0.20
1.1.17	Margarine and other vegetable fats	0.30	0.30	0.40	0.40	0.40	0.40
1.1.18	Peanut butter	[0.00]	[0.00]	[0.00]	0.00	0.00	0.00
1.1.19	Cooking oils and fats	0.10	0.10	0.10	0.20	0.20	0.20
1.1.20	Fresh fruit	1.20	1.50	1.90	1.90	2.20	2.30
1.1.21	Other fresh, chilled or frozen fruits	0.10	0.10	0.10	0.20	0.20	0.20
1.1.22	Dried fruit and nuts	0.10	0.20	0.30	0.20	0.40	0.40
1.1.23	Preserved fruit and fruit based products	0.10	0.10	0.20	0.10	0.20	0.10
1.1.24	Fresh vegetables	1.40	1.70	2.30	2.30	2.60	2.80
1.1.25	Dried vegetables and other preserved or processed vegetables	0.50	0.60	0.80	0.80	0.90	1.00
1.1.26	Potatoes	0.40	0.60	0.70	0.60	0.70	0.70
1.1.27	Other tubers and products of tuber vegetables	0.50	0.70	0.90	1.10	1.20	1.30
1.1.28	Sugar and sugar products	0.20	0.20	0.30	0.30	0.30	0.30
1.1.29	Jams, marmalades	0.10	0.20	0.20	0.20	0.20	0.20
1.1.30	Chocolate	0.60	0.70	1.00	1.20	1.20	1.30
1.1.31	Confectionery products	0.30	0.40	0.50	0.60	0.60	0.60
1.1.32	Edible ices and ice cream	0.20	0.30	0.40	0.50	0.40	0.40
1.1.33	Other food products	0.80	1.00	1.50	1.40	1.60	1.90
1.2	Non-alcoholic drinks	1.80	2.20	2.80	3.10	3.50	3.60
1.2.1	Coffee	0.30	0.30	0.50	0.40	0.50	0.50
1.2.2	Tea	0.30	0.40	0.50	0.40	0.50	0.50
1.2.3	Cocoa and powdered chocolate	0.10	0.10	0.10	0.10	0.10	0.10
1.2.4	Fruit and vegetable juices, mineral waters	0.50	0.60	0.70	0.80	1.00	1.10
1.2.5	Soft drinks	0.60	0.80	1.00	1.30	1.50	1.50

Note: The commodity and service categories are not comparable with those in publications before 2001-02
The numbering system is sequential, it does not use actual COICOP codes

1.3 Detailed household expenditure by gross income decile group (cont.) 2002-03

based on weighted data and including children's expenditure

			Seventh decile group	Eighth decile group	Ninth decile group	Highest ten per cent	All house-holds
Lower boundary of group (£ per week)			541	662	821	1,085	
Grossed number of households (thousands)			2,430	2,440	2,430	2,430	24,350
Total number of households in sample			687	672	655	625	6,927
Total number of persons in sample			1,967	1,917	1,974	2,022	16,586
Total number of adults in sample			1,409	1,407	1,477	1,495	12,450
Weighted average number of persons per household			2.8	2.8	3.0	3.2	2.4
Commodity or service			**Average weekly household expenditure (£)**				
1	**Food & non-alcoholic drinks**		**49.00**	**52.00**	**56.90**	**66.90**	**42.70**
1.1	Food		44.70	47.30	51.90	61.20	39.10
	1.1.1	Bread, rice and cereals	4.40	4.80	5.10	5.70	3.90
	1.1.2	Pasta products	0.40	0.40	0.50	0.60	0.30
	1.1.3	Buns, cakes, biscuits etc.	3.30	3.20	3.50	4.00	2.80
	1.1.4	Pastry (savoury)	0.70	0.90	1.00	1.10	0.60
	1.1.5	Beef (fresh, chilled or frozen)	1.60	1.70	1.80	2.00	1.40
	1.1.6	Pork (fresh, chilled or frozen)	0.80	0.80	0.70	0.70	0.60
	1.1.7	Lamb (fresh, chilled or frozen)	0.60	0.70	0.70	1.00	0.60
	1.1.8	Poultry (fresh, chilled or frozen)	1.70	1.90	2.40	2.40	1.50
	1.1.9	Bacon and ham	1.00	1.00	1.00	1.20	0.90
	1.1.10	Other meat and meat preparations	5.60	5.60	6.60	7.30	4.80
	1.1.11	Fish and fish products	1.80	2.20	2.50	3.00	1.80
	1.1.12	Milk	2.60	2.20	2.40	2.50	2.10
	1.1.13	Cheese and curd	1.50	1.80	2.00	2.50	1.40
	1.1.14	Eggs	0.40	0.40	0.50	0.60	0.40
	1.1.15	Other milk products	1.70	1.80	1.90	2.40	1.40
	1.1.16	Butter	0.20	0.30	0.30	0.40	0.30
	1.1.17	Margarine and other vegetable fats	0.50	0.50	0.40	0.50	0.40
	1.1.18	Peanut butter	0.00	0.00	0.00	0.00	0.00
	1.1.19	Cooking oils and fats	0.20	0.30	0.20	0.30	0.20
	1.1.20	Fresh fruit	2.70	3.00	3.20	4.20	2.40
	1.1.21	Other fresh, chilled or frozen fruits	0.20	0.30	0.30	0.50	0.20
	1.1.22	Dried fruit and nuts	0.40	0.40	0.40	0.70	0.30
	1.1.23	Preserved fruit and fruit based products	0.10	0.20	0.10	0.20	0.10
	1.1.24	Fresh vegetables	3.50	3.80	4.20	5.70	3.00
	1.1.25	Dried vegetables and other preserved or processed vegetables	1.20	1.30	1.50	1.80	1.00
	1.1.26	Potatoes	0.90	0.90	0.90	1.00	0.70
	1.1.27	Other tubers and products of tuber vegetables	1.50	1.40	1.50	1.50	1.20
	1.1.28	Sugar and sugar products	0.30	0.30	0.20	0.30	0.30
	1.1.29	Jams, marmalades	0.20	0.20	0.20	0.30	0.20
	1.1.30	Chocolate	1.50	1.50	1.80	1.70	1.20
	1.1.31	Confectionery products	0.70	0.60	0.70	0.70	0.60
	1.1.32	Edible ices and ice cream	0.50	0.60	0.60	0.90	0.50
	1.1.33	Other food products	2.00	2.40	2.60	3.10	1.80
1.2	Non-alcoholic drinks		4.30	4.70	5.00	5.70	3.70
	1.2.1	Coffee	0.60	0.70	0.60	0.60	0.50
	1.2.2	Tea	0.50	0.40	0.60	0.50	0.50
	1.2.3	Cocoa and powdered chocolate	0.10	0.10	0.10	0.10	0.10
	1.2.4	Fruit and vegetable juices, mineral waters	1.20	1.50	1.70	2.30	1.20
	1.2.5	Soft drinks	1.80	1.90	2.10	2.00	1.50

Note: The commodity and service categories are not comparable with those in publications before 2001-02
The numbering system is sequential, it does not use actual COICOP codes

1.3 Detailed household expenditure by gross income decile group (cont.) 2002-03

based on weighted data and including children's expenditure

	Commodity or service	Lowest ten per cent	Second decile group	Third decile group	Fourth decile group	Fifth decile group	Sixth decile group
		Average weekly household expenditure (£)					
2	**Alcoholic drink, tobacco & narcotics**	**5.90**	**6.30**	**8.80**	**10.50**	**10.60**	**12.00**
2.1	Alcoholic drinks	2.10	2.50	3.70	4.30	5.10	5.60
2.1.1	Spirits and liqueurs (brought home)	0.60	0.70	1.30	1.00	1.20	1.30
2.1.2	Wines, fortified wines (brought home)	0.80	1.00	1.30	1.70	2.30	2.40
2.1.3	Beer, lager, ciders and Perry (brought home)	0.70	0.70	1.10	1.60	1.60	1.70
2.1.4	Alcopops (brought home)	..	[0.10]	[0.10]	0.10	0.10	0.20
2.2	Tobacco and narcotics	3.80	3.80	5.00	6.20	5.50	6.50
2.2.1	Cigarettes	3.30	3.40	4.60	5.60	4.60	5.80
2.2.2	Cigars, other tobacco products and narcotics	0.50	0.50	0.40	0.60	0.90	0.70
3	**Clothing & footwear**	**6.20**	**7.90**	**11.40**	**15.80**	**18.00**	**20.20**
3.1	Clothing	5.00	6.20	9.20	12.80	14.30	16.90
3.1.1	Men's outer garments	1.00	0.80	1.90	2.90	2.60	3.90
3.1.2	Men's under garments	0.20	0.30	0.30	0.20	0.50	0.30
3.1.3	Women's outer garments	2.20	2.40	4.00	5.50	6.70	7.40
3.1.4	Women's under garments	0.50	0.60	0.80	1.10	1.00	1.00
3.1.5	Boys' outer garments (5-15)	[0.20]	0.50	0.50	0.70	0.80	0.80
3.1.6	Girls' outer garments (5-15)	[0.20]	0.60	0.70	1.00	1.00	1.30
3.1.7	Infants' outer garments (under 5)	0.10	0.30	0.30	0.40	0.50	0.70
3.1.8	Children's under garments (under 16)	0.10	0.20	0.30	0.40	0.40	0.40
3.1.9	Accessories	0.20	0.20	0.20	0.40	0.40	0.60
3.1.10	Haberdashery and clothing hire	0.10	0.10	0.10	0.10	0.20	0.20
3.1.11	Dry cleaners, laundry and dyeing	0.20	[0.10]	0.20	[0.10]	0.20	0.10
3.2	Footwear	1.20	1.70	2.20	3.10	3.70	3.30
4	**Housing (net)[1], fuel & power**	**20.90**	**28.40**	**29.30**	**34.80**	**37.60**	**38.10**
4.1	Actual rentals for housing	42.60	37.40	24.60	22.10	19.30	15.50
4.1.1	Gross rent	42.60	37.40	24.50	22.10	19.30	15.50
4.1.2	*less* housing benefit, rebates & allowances rec'd	36.10	27.50	14.20	8.70	3.70	1.90
4.1.3	Net rent	6.40	9.90	10.40	13.50	15.60	13.60
4.1.4	Second dwelling rent	..	..	..	..	..	..
4.2	Maintenance and repair of dwelling	2.20	3.90	4.10	5.40	5.60	7.20
4.3	Water supply and miscellaneous services relating to the dwelling	4.50	5.10	4.80	5.20	5.30	5.30
4.4	Electricity, gas and other fuels	7.70	9.40	10.00	10.80	11.10	12.00
4.4.1	Electricity	4.20	4.70	5.10	5.40	5.70	5.90
4.4.2	Gas	3.10	3.90	4.20	4.70	4.70	5.50
4.4.3	Other fuels	0.40	0.80	0.60	0.70	0.60	0.60

Note: The commodity and service categories are not comparable with those in publications before 2001-02
The numbering system is sequential, it does not use actual COICOP codes
1 Excluding mortgage interest payments, council tax and Northern Ireland rates

1.3 Detailed household expenditure by gross income decile group (cont.) 2002-03

based on weighted data and including children's expenditure

	Commodity or service	Seventh decile group	Eighth decile group	Ninth decile group	Highest ten per cent	All house-holds
		Average weekly household expenditure (£)				
2	**Alcoholic drink, tobacco & narcotics**	**14.10**	**13.10**	**15.80**	**16.40**	**11.40**
2.1	Alcoholic drinks	7.20	7.70	9.80	11.40	5.90
2.1.1	Spirits and liqueurs (brought home)	1.50	1.30	1.60	1.80	1.20
2.1.2	Wines, fortified wines (brought home)	3.30	3.90	5.30	6.90	2.90
2.1.3	Beer, lager, ciders and Perry (brought home)	2.10	2.40	2.60	2.50	1.70
2.1.4	Alcopops (brought home)	0.30	0.20	0.20	0.20	0.10
2.2	Tobacco and narcotics	7.00	5.40	6.00	5.10	5.40
2.2.1	Cigarettes	6.30	4.80	5.00	4.60	4.80
2.2.2	Cigars, other tobacco products and narcotics	0.70	0.60	1.00	0.50	0.60
3	**Clothing & footwear**	**26.20**	**31.00**	**36.60**	**49.60**	**22.30**
3.1	Clothing	21.30	25.60	30.50	41.40	18.30
3.1.1	Men's outer garments	5.60	6.30	8.10	11.60	4.50
3.1.2	Men's under garments	0.60	0.50	0.60	0.90	0.40
3.1.3	Women's outer garments	8.60	11.60	13.60	18.20	8.00
3.1.4	Women's under garments	1.50	1.60	2.00	2.60	1.30
3.1.5	Boys' outer garments (5-15)	1.20	1.10	1.30	1.10	0.80
3.1.6	Girls' outer garments (5-15)	1.40	1.60	1.40	2.00	1.10
3.1.7	Infants' outer garments (under 5)	0.80	0.90	0.80	1.10	0.60
3.1.8	Children's under garments (under 16)	0.50	0.50	0.50	0.60	0.40
3.1.9	Accessories	0.70	1.00	1.20	1.90	0.70
3.1.10	Haberdashery and clothing hire	0.20	0.20	0.60	0.30	0.20
3.1.11	Dry cleaners, laundry and dyeing	0.30	0.30	0.50	1.00	0.30
3.2	Footwear	4.90	5.40	6.00	8.30	4.00
4	**Housing (net)[1], fuel & power**	**37.10**	**39.70**	**42.80**	**60.00**	**36.90**
4.1	Actual rentals for housing	13.10	11.50	12.80	17.30	21.60
4.1.1	Gross rent	13.10	11.40	12.70	17.30	21.60
4.1.2	*less* housing benefit, rebates & allowances rec'd	[0.70]	[0.60]	..	..	9.40
4.1.3	Net rent	12.50	10.80	12.10	17.20	12.20
4.1.4	Second dwelling rent	..	..	..	..	..
4.2	Maintenance and repair of dwelling	7.00	9.90	10.20	19.80	7.50
4.3	Water supply and miscellaneous services relating to the dwelling	5.30	5.60	6.60	6.40	5.40
4.4	Electricity, gas and other fuels	12.30	13.20	13.80	16.60	11.70
4.4.1	Electricity	6.10	6.30	6.80	7.70	5.80
4.4.2	Gas	5.40	5.90	6.30	7.60	5.10
4.4.3	Other fuels	0.90	1.10	0.80	1.30	0.80

Note: The commodity and service categories are not comparable with those in publications before 2001-02
The numbering system is sequential, it does not use actual COICOP codes
1 Excluding mortgage interest payments, council tax and Northern Ireland rates

1.3 Detailed household expenditure by gross income decile group (cont.) 2002-03

based on weighted data and including children's expenditure

		Lowest ten per cent	Second decile group	Third decile group	Fourth decile group	Fifth decile group	Sixth decile group
Commodity or service		Average weekly household expenditure (£)					
5	**Household goods & services**	**11.00**	**12.70**	**17.60**	**23.20**	**22.00**	**29.80**
5.1	Furniture and furnishings, carpets and other floor coverings	5.10	6.70	7.10	11.10	11.60	16.80
5.1.1	Furniture and furnishings	3.20	4.90	5.10	7.90	9.30	12.30
5.1.2	Floor coverings	1.80	1.70	2.00	3.20	2.30	4.50
5.1.3	Repair of furniture, furnishings and floor coverings	..	..	..	..	..	..
5.2	Household textiles	0.90	0.60	1.30	1.80	1.40	1.70
5.3	Household appliances	1.70	1.10	2.80	3.70	2.10	3.50
5.4	Glassware, tableware and household utensils	0.50	0.60	0.80	1.10	0.90	1.40
5.5	Tools and equipment for house and garden	0.80	0.90	2.20	1.80	2.60	2.40
5.6	Goods and services for routine household maintenance	2.00	2.80	3.50	3.60	3.50	3.90
5.6.1	Cleaning materials	0.90	1.40	1.60	1.70	2.00	2.10
5.6.2	Household goods and hardware	0.50	0.50	0.80	0.90	1.00	1.00
5.6.3	Domestic services, carpet cleaning	0.60	0.90	1.00	1.10	0.50	0.80
6	**Health**	**1.60**	**2.20**	**4.40**	**3.10**	**4.50**	**5.50**
6.1	Medical products, appliances and equipment	1.30	1.50	3.50	2.20	3.50	2.50
6.1.1	Medicines, prescriptions and healthcare products	0.70	0.80	1.20	1.10	1.40	1.60
6.1.2	Spectacles, lenses, accessories and repairs	[0.60]	[0.60]	1.90	1.00	1.90	0.90
6.1.3	Non-optical appliances and equipment (e.g. wheelchairs, batteries for hearing aids, etc.)	..	..	..	..	..	..
6.2	Hospital services	[0.30]	0.80	0.90	0.90	1.10	3.00
7	**Transport**	**13.10**	**16.10**	**22.60**	**36.90**	**46.90**	**58.10**
7.1	Purchase of vehicles	5.20	5.10	7.10	14.90	22.00	26.40
7.1.1	Purchase of new cars and vans	..	..	[2.50]	4.70	10.40	10.50
7.1.2	Purchase of second hand cars or vans	3.70	2.80	4.40	10.00	11.10	15.10
7.1.3	Purchase of motorcycles and other vehicles	..	..	..	..	..	[0.70]
7.2	Operation of personal transport	5.00	7.50	11.40	17.30	19.00	25.30
7.2.1	Spares and accessories	[0.40]	0.70	1.40	1.40	1.10	2.30
7.2.2	Petrol, diesel and other motor oils	3.00	4.30	6.50	10.70	12.30	15.60
7.2.3	Repairs and servicing	1.20	1.80	2.40	3.70	4.20	5.80
7.2.4	Other motoring costs	0.40	0.70	1.10	1.50	1.40	1.60
7.3	Transport services	2.90	3.50	4.10	4.60	5.90	6.50
7.3.1	Rail and tube fares	0.40	0.40	0.60	0.50	1.00	1.40
7.3.2	Bus and coach fares	1.00	0.90	1.40	1.20	1.50	1.60
7.3.3	Combined fares	..	..	..	[0.30]	[0.60]	[0.30]
7.3.4	Other travel and transport	1.10	2.10	2.20	2.50	2.80	3.20

Note: The commodity and service categories are not comparable with those in publications before 2001-02
The numbering system is sequential, it does not use actual COICOP codes

1.3 Detailed household expenditure by gross income decile group (cont.) 2002-03

based on weighted data and including children's expenditure

Commodity or service		Seventh decile group	Eighth decile group	Ninth decile group	Highest ten per cent	All house-holds
		Average weekly household expenditure (£)				
5	**Household goods & services**	**33.00**	**36.70**	**44.40**	**71.90**	**30.20**
5.1	Furniture and furnishings, carpets and other floor coverings	16.90	20.80	24.80	39.60	16.10
5.1.1	Furniture and furnishings	13.20	16.90	19.10	33.30	12.50
5.1.2	Floor coverings	3.80	3.80	5.80	6.30	3.50
5.1.3	Repair of furniture, furnishings and floor coverings	..	..	..	..	..
5.2	Household textiles	2.90	2.20	3.10	4.70	2.10
5.3	Household appliances	3.40	3.40	4.10	6.60	3.20
5.4	Glassware, tableware and household utensils	1.60	2.20	2.50	4.80	1.60
5.5	Tools and equipment for house and garden	3.50	2.70	4.40	5.30	2.70
5.6	Goods and services for routine household maintenance	4.80	5.40	5.50	10.90	4.60
5.6.1	Cleaning materials	2.30	2.60	2.60	3.10	2.00
5.6.2	Household goods and hardware	1.40	1.30	1.40	1.80	1.10
5.6.3	Domestic services, carpet cleaning	1.00	1.50	1.40	6.00	1.50
6	**Health**	**5.20**	**6.50**	**5.30**	**9.90**	**4.80**
6.1	Medical products, appliances and equipment	3.10	4.00	4.00	6.20	3.20
6.1.1	Medicines, prescriptions and healthcare products	1.60	2.10	1.90	2.50	1.50
6.1.2	Spectacles, lenses, accessories and repairs	1.50	1.80	2.00	3.70	1.60
6.1.3	Non-optical appliances and equipment (e.g. wheelchairs, batteries for hearing aids, etc.)	..	..	..	..	0.10
6.2	Hospital services	2.10	2.60	1.30	3.70	1.70
7	**Transport**	**66.30**	**84.80**	**107.80**	**139.60**	**59.20**
7.1	Purchase of vehicles	29.50	37.40	51.60	67.10	26.60
7.1.1	Purchase of new cars and vans	8.50	16.20	20.70	35.70	11.30
7.1.2	Purchase of second hand cars or vans	19.40	20.20	28.50	29.60	14.50
7.1.3	Purchase of motorcycles and other vehicles	[1.60]	0.90	2.40	1.70	0.90
7.2	Operation of personal transport	28.20	37.30	42.00	47.60	24.10
7.2.1	Spares and accessories	2.10	4.40	4.60	2.80	2.10
7.2.2	Petrol, diesel and other motor oils	18.20	22.30	26.60	28.80	14.80
7.2.3	Repairs and servicing	5.40	8.10	7.50	11.70	5.20
7.2.4	Other motoring costs	2.60	2.60	3.40	4.30	1.90
7.3	Transport services	8.60	10.10	14.20	24.90	8.50
7.3.1	Rail and tube fares	1.70	1.80	3.30	7.20	1.80
7.3.2	Bus and coach fares	1.30	1.30	1.60	2.30	1.40
7.3.3	Combined fares	[0.60]	1.10	1.40	3.20	0.80
7.3.4	Other travel and transport	5.00	5.80	7.90	12.10	4.50

Note: The commodity and service categories are not comparable with those in publications before 2001-02
The numbering system is sequential, it does not use actual COICOP codes

1.3 Detailed household expenditure by gross income decile group (cont.) 2002-03

based on weighted data and including children's expenditure

		Lowest ten per cent	Second decile group	Third decile group	Fourth decile group	Fifth decile group	Sixth decile group
Commodity or service		Average weekly household expenditure (£)					
8	**Communication**	**5.40**	**5.90**	**7.50**	**8.80**	**9.50**	**10.70**
8.1	Postal services	0.40	0.30	0.40	0.40	0.50	0.40
8.2	Telephone and telefax equipment	..	..	[0.40]	0.60	0.30	[0.60]
8.3	Telephone and telefax services	5.00	5.50	6.70	7.90	8.70	9.70
9	**Recreation & culture**	**16.60**	**21.30**	**28.60**	**43.10**	**46.70**	**54.70**
9.1	Audio-visual, photographic and information processing equipment	2.10	1.70	3.20	6.30	7.60	7.30
9.1.1	Audio equipment and accessories, CD players	0.80	0.50	0.80	1.90	2.20	2.30
9.1.2	TV, video and computers	1.30	1.10	2.20	4.10	4.40	4.00
9.1.3	Photographic, cinematographic & optical equip't	..	[0.10]	0.20	0.30	1.00	1.10
9.2	Other major durables for recreation and culture	..	..	..	[1.30]	[0.50]	..
9.3	Other recreational items and equipment, gardens and pets	2.60	4.10	5.00	7.40	8.80	9.00
9.3.1	Games, toys and hobbies	0.70	0.70	1.00	1.70	2.20	2.10
9.3.2	Computer software and games	..	[0.40]	0.90	1.10	0.90	1.10
9.3.3	Equipment for sport, camping and open-air recreation	[0.10]	0.20	0.30	0.30	0.50	0.50
9.3.4	Horticultural goods, garden equipment and plants	0.80	1.20	1.30	1.60	2.60	2.70
9.3.5	Pets and pet food	0.90	1.60	1.60	2.70	2.60	2.70
9.4	Recreational and cultural services	6.50	7.40	9.40	12.60	13.60	15.90
9.4.1	Sports admissions, subscriptions and leisure class fees	1.50	0.90	1.60	2.10	2.80	4.10
9.4.2	Cinema, theatre and museums etc.	0.30	0.40	0.60	1.00	1.10	1.60
9.4.3	TV, video, satellite rental, cable subscriptions, TV licences and the Internet	2.60	2.80	3.30	4.40	4.80	5.40
9.4.4	Miscellaneous entertainments	0.30	0.30	0.50	0.80	0.90	0.90
9.4.5	Development of film, deposit for film development, passport photos, holiday and school photos	[0.10]	0.10	0.30	0.20	0.30	0.40
9.4.6	Gambling payments	1.70	2.90	3.20	4.10	3.70	3.50
9.5	Newspapers, books and stationery	2.80	4.00	4.30	5.30	6.20	5.50
9.5.1	Books, diaries, address books, cards etc.	1.20	1.90	1.70	2.70	3.20	2.90
9.5.2	Newspapers	1.20	1.60	1.90	1.80	2.00	1.70
9.5.3	Magazines and periodicals	0.50	0.60	0.70	0.90	1.00	0.90
9.6	Package holidays	2.30	4.00	6.50	10.20	10.00	11.40
9.6.1	Package holidays - UK	0.80	0.60	1.50	0.80	0.60	1.20
9.6.2	Package holidays - abroad	1.50	3.40	4.90	9.40	9.50	10.20
10	**Education**	**[0.80]**	**1.70**	**0.70**	**1.20**	**2.00**	**3.60**
10.1	Education fees	[0.80]	[1.60]	[0.40]	1.00	1.80	3.30
10.2	Payments for school trips, other ad-hoc expenditure	..	[0.10]	[0.20]	[0.10]	[0.10]	[0.30]

Note: The commodity and service categories are not comparable with those in publications before 2001-02
The numbering system is sequential, it does not use actual COICOP codes

1.3 Detailed household expenditure by gross income decile group (cont.) 2002-03

based on weighted data and including children's expenditure

		Seventh decile group	Eighth decile group	Ninth decile group	Highest ten per cent	All house-holds
Commodity or service		Average weekly household expenditure (£)				
8	**Communication**	**11.30**	**13.60**	**15.00**	**18.30**	**10.60**
8.1	Postal services	0.50	0.70	0.60	0.70	0.50
8.2	Telephone and telefax equipment	0.90	1.10	1.10	1.20	0.60
8.3	Telephone and telefax services	9.90	11.80	13.30	16.40	9.50
9	**Recreation & culture**	**68.40**	**79.00**	**89.30**	**116.40**	**56.40**
9.1	Audio-visual, photographic and information processing equipment	12.60	11.30	14.00	15.50	8.20
9.1.1	Audio equipment and accessories, CD players	2.90	3.10	3.80	4.20	2.30
9.1.2	TV, video and computers	8.20	7.70	9.20	8.10	5.00
9.1.3	Photographic, cinematographic & optical equip't	1.50	0.50	0.90	3.10	0.90
9.2	Other major durables for recreation and culture	[1.60]	..	2.10	2.50	1.80
9.3	Other recreational items and equipment, gardens and pets	10.90	13.80	15.10	22.80	10.00
9.3.1	Games, toys and hobbies	2.80	2.90	3.80	4.40	2.20
9.3.2	Computer software and games	1.20	2.20	1.00	1.60	1.10
9.3.3	Equipment for sport, camping and open-air recreation	1.10	1.10	1.90	2.30	0.80
9.3.4	Horticultural goods, garden equipment and plants	3.10	3.70	4.40	8.60	3.00
9.3.5	Pets and pet food	2.70	3.80	4.00	5.90	2.90
9.4	Recreational and cultural services	19.60	24.30	26.60	36.60	17.20
9.4.1	Sports admissions, subscriptions and leisure class fees	5.60	10.10	8.80	17.90	5.60
9.4.2	Cinema, theatre and museums etc.	1.90	2.20	2.70	4.60	1.60
9.4.3	TV, video, satellite rental, cable subscriptions, TV licences and the Internet	5.50	6.10	6.80	6.90	4.80
9.4.4	Miscellaneous entertainments	1.40	1.30	1.70	2.20	1.00
9.4.5	Development of film, deposit for film development, passport photos, holiday and school photos	0.50	0.50	1.50	0.70	0.50
9.4.6	Gambling payments	4.70	4.20	5.00	4.30	3.70
9.5	Newspapers, books and stationery	7.30	8.40	9.30	12.60	6.60
9.5.1	Books, diaries, address books, cards etc.	4.00	4.80	5.40	8.00	3.60
9.5.2	Newspapers	1.90	2.10	2.30	2.60	1.90
9.5.3	Magazines and periodicals	1.40	1.50	1.60	2.00	1.10
9.6	Package holidays	16.30	17.40	22.20	26.40	12.70
9.6.1	Package holidays - UK	[1.00]	[0.70]	[1.10]	[0.80]	0.90
9.6.2	Package holidays - abroad	15.30	16.60	21.10	25.60	11.70
10	**Education**	**3.40**	**4.30**	**8.50**	**25.80**	**5.20**
10.1	Education fees	3.00	4.10	8.10	25.00	4.90
10.2	Payments for school trips, other ad-hoc expenditure	0.50	0.20	0.40	0.90	0.30

Note: The commodity and service categories are not comparable with those in publications before 2001-02
The numbering system is sequential, it does not use actual COICOP codes

1.3 Detailed household expenditure by gross income decile group (cont.) 2002-03

based on weighted data and including children's expenditure

Commodity or service	Lowest ten per cent	Second decile group	Third decile group	Fourth decile group	Fifth decile group	Sixth decile group
	Average weekly household expenditure (£)					
11 Restaurants & hotels	**10.20**	**11.60**	**15.50**	**22.30**	**28.50**	**32.50**
11.1 Catering services	8.40	10.00	13.50	19.80	25.20	28.70
11.1.1 Restaurant and café meals	3.30	3.80	5.10	7.10	10.00	10.00
11.1.2 Alcoholic drinks (away from home)	2.50	2.70	4.00	5.70	7.10	8.70
11.1.3 Take away meals eaten at home	1.40	1.80	2.10	2.80	2.90	3.80
11.1.4 Other take-away and snack food	0.90	1.40	1.90	3.20	3.60	4.30
11.1.5 Contract catering (food)	..	..	..	..	..	..
11.1.6 Canteens	0.20	0.20	0.50	1.00	1.50	1.80
11.2 Accommodation services	1.80	1.60	2.00	2.50	3.20	3.80
11.2.1 Holiday in the UK	0.70	1.00	1.30	1.50	2.40	2.50
11.2.2 Holiday abroad	..	[0.60]	[0.60]	1.00	0.80	1.40
11.2.3 Room hire	..	..	..	..	..	..
12 Miscellaneous goods & services	**10.20**	**12.70**	**16.00**	**22.10**	**25.90**	**33.80**
12.1 Personal care	3.10	4.00	5.30	6.60	7.40	9.50
12.1.1 Hairdressing, beauty treatment	1.10	1.30	1.80	1.90	2.00	3.00
12.1.2 Toilet paper	0.30	0.40	0.60	0.60	0.70	0.70
12.1.3 Toiletries and soap	0.70	0.80	1.20	1.40	1.80	2.00
12.1.4 Baby toiletries and accessories (disposable)	0.10	0.40	0.30	0.40	0.50	0.70
12.1.5 Hair products, cosmetics and related electrical appliances	0.80	1.10	1.40	2.20	2.50	3.10
12.2 Personal effects	0.60	0.90	0.90	1.90	1.90	2.70
12.3 Social protection	[0.40]	0.60	1.00	0.80	1.10	1.70
12.4 Insurance	4.60	5.70	7.80	10.50	12.40	14.90
12.4.1 Household insurances - structural, contents and appliances	1.80	2.20	3.10	3.60	3.90	4.60
12.4.2 Medical insurance premiums	..	[0.10]	0.30	1.00	1.00	1.30
12.4.3 Vehicle insurance including boat insurance	2.60	3.30	4.20	5.80	7.30	8.80
12.4.4 Non-package holiday, other travel insurance	..	..	..	..	..	..
12.5 Other services n.e.c	1.40	1.60	1.00	2.30	2.90	5.10
12.5.1 Moving house	0.90	0.80	0.60	1.50	1.80	2.90
12.5.2 Bank, building society, post office, credit card charges	0.10	0.10	0.10	0.30	0.30	0.40
12.5.3 Other services and professional fees	0.40	0.60	0.30	0.60	0.90	1.70
1-12 All expenditure groups	**123.10**	**154.50**	**196.90**	**258.30**	**292.30**	**341.80**

Note: The commodity and service categories are not comparable with those in publications before 2001-02
The numbering system is sequential, it does not use actual COICOP codes

1.3 Detailed household expenditure by gross income decile group (cont.) 2002-03

based on weighted data and including children's expenditure

Commodity or service	Seventh decile group	Eighth decile group	Ninth decile group	Highest ten per cent	All households
	Average weekly household expenditure (£)				
11 Restaurants & hotels	**40.80**	**48.10**	**61.50**	**83.40**	**35.40**
11.1 Catering services	35.90	41.70	52.70	68.80	30.50
11.1.1 Restaurant and café meals	11.60	15.60	18.60	27.60	11.30
11.1.2 Alcoholic drinks (away from home)	11.70	12.10	15.50	18.80	8.90
11.1.3 Take away meals eaten at home	4.40	4.60	6.00	5.90	3.60
11.1.4 Other take-away and snack food	5.40	5.80	7.40	8.30	4.20
11.1.5 Contract catering (food)	..	[0.70]	..	..	0.70
11.1.6 Canteens	2.70	2.80	3.70	4.10	1.80
11.2 Accommodation services	4.90	6.40	8.80	14.70	5.00
11.2.1 Holiday in the UK	3.20	3.30	4.50	4.50	2.50
11.2.2 Holiday abroad	1.60	3.00	4.20	10.10	2.40
11.2.3 Room hire	..	..	..	..	0.00
12 Miscellaneous goods & services	**37.40**	**45.10**	**51.20**	**76.50**	**33.10**
12.1 Personal care	9.90	11.40	13.20	16.60	8.70
12.1.1 Hairdressing, beauty treatment	2.60	3.50	4.20	5.60	2.70
12.1.2 Toilet paper	0.80	0.80	0.80	0.90	0.70
12.1.3 Toiletries and soap	2.20	2.70	2.70	3.50	1.90
12.1.4 Baby toiletries and accessories (disposable)	0.70	0.60	0.70	1.00	0.50
12.1.5 Hair products, cosmetics and electrical personal appliances	3.70	3.80	4.70	5.60	2.90
12.2 Personal effects	3.20	4.10	3.90	8.30	2.80
12.3 Social protection	2.10	3.40	4.80	10.30	2.60
12.4 Insurance	17.60	19.30	24.80	29.20	14.70
12.4.1 Household insurances - structural, contents and appliances	5.30	5.60	6.30	8.70	4.50
12.4.2 Medical insurance premiums	1.30	2.20	2.50	3.90	1.40
12.4.3 Vehicle insurance including boat insurance	10.80	11.20	15.40	16.20	8.60
12.4.4 Non-package holiday, other travel insurance	..	[0.20]	[0.60]	[0.40]	0.20
12.5 Other services	4.60	6.90	4.50	12.10	4.30
12.5.1 Moving house	2.80	3.50	2.20	7.50	2.40
12.6.2 Bank, building society, post office, credit card charges	0.40	0.50	0.60	0.80	0.40
12.6.3 Other services and professional fees	1.40	2.90	1.80	3.90	1.40
1-12 All expenditure groups	**392.30**	**454.10**	**535.20**	**734.90**	**348.30**

Note: The commodity and service categories are not comparable with those in publications before 2001-02
The numbering system is sequential, it does not use actual COICOP codes

1.3 Detailed household expenditure by gross income decile group (cont.) 2002-03

based on weighted data and including children's expenditure

Commodity or service	Lowest ten per cent	Second decile group	Third decile group	Fourth decile group	Fifth decile group	Sixth decile group
	Average weekly household expenditure (£)					
13 Other expenditure items	**12.50**	**14.90**	**22.50**	**36.90**	**42.70**	**58.60**
13.1 Housing: mortgage interest payments, water, council tax etc.	7.70	9.10	14.20	24.40	29.40	41.00
13.2 Licences, fines and transfers	0.80	1.00	1.40	2.00	3.00	3.00
13.3 Holiday spending	[1.20]	[1.50]	[2.00]	4.20	2.80	6.10
13.4 Money transfers and credit	2.80	3.30	4.80	6.30	7.60	8.60
13.4.1 Money, cash gifts given to children	..	[0.00]	..	[0.10]	..	0.10
13.4.2 Cash gifts and donations	2.50	2.80	4.20	5.00	5.80	6.40
13.4.3 Club instalment payments (child) and interest on credit cards	0.30	0.40	0.50	1.20	1.80	2.00
Total expenditure	**135.60**	**169.40**	**219.40**	**295.20**	**335.00**	**400.40**
14 Other items recorded						
14.1 Life assurance and contributions to pension funds	1.90	1.90	4.70	7.90	11.00	17.30
14.2 Other insurance inc. Friendly Societies	0.20	0.40	0.50	0.80	1.00	1.10
14.3 Income tax, payments less refunds	0.60	3.30	8.90	20.40	34.30	51.20
14.4 National insurance contributions	[0.10]	0.40	1.70	5.80	10.70	17.20
14.5 Purchase or alteration of dwellings, mortgages	3.50	5.40	8.60	14.60	19.60	25.50
14.6 Savings and investments	0.40	0.60	1.00	1.40	2.60	3.90
14.7 Pay off loan to clear other debt	[0.30]	..	0.70	1.90	2.40	3.30
14.8 Windfall receipts from gambling etc.	0.70	1.30	2.20	2.60	1.80	1.10

Note: The commodity and service categories are not comparable with those in publications before 2001-02
The numbering system is sequential, it does not use actual COICOP codes

1.3 Detailed household expenditure by gross income decile group (cont.) 2002-03

based on weighted data and including children's expenditure

	Seventh decile group	Eighth decile group	Ninth decile group	Highest ten per cent	All households
Commodity or service	**Average weekly household expenditure (£)**				
13 Other expenditure items	**63.60**	**80.00**	**99.00**	**148.50**	**57.90**
13.1 Housing: mortgage interest payments, water, council tax etc.	47.40	58.00	65.50	97.60	39.40
13.2 Licences, fines and transfers	3.20	3.70	4.30	5.10	2.70
13.3 Holiday spending	4.50	6.80	10.30	24.70	6.40
13.4 Money transfers and credit	8.50	11.50	19.00	21.00	9.40
13.4.1 Money, cash gifts given to children	0.10	0.20	..	0.20	0.20
13.4.2 Cash gifts and donations	6.40	8.90	15.30	18.00	7.50
13.4.3 Club instalment payments (child) and interest on credit cards	2.00	2.40	3.00	2.80	1.60
Total expenditure	**455.90**	**534.10**	**634.20**	**883.40**	**406.20**
14 Other items recorded					
14.1 Contributions to pension funds	22.40	31.40	41.00	90.80	23.00
14.2 Other insurance inc. Friendly Societies	1.40	1.80	2.50	2.80	1.20
14.3 Income tax, payments less refunds	68.20	96.00	136.00	318.60	73.70
14.4 National insurance contributions	25.70	33.70	41.90	56.40	19.40
14.5 Purchase or alteration of dwellings, mortgages	30.60	44.60	61.70	105.30	31.90
14.6 Savings and investments	6.10	8.40	12.20	27.00	6.40
14.7 Pay off loan to clear other debt	3.60	3.40	5.20	3.00	2.50
14.8 Windfall receipts from gambling etc.	3.60	2.70	1.90	2.30	2.00

Note: The commodity and service categories are not comparable with those in publications before 2001-02
The numbering system is sequential, it does not use actual COICOP codes

1.4 Household expenditure by disposable income decile group 2002-03

based on weighted data and including children's expenditure

	Lowest ten per cent	Second decile group	Third decile group	Fourth decile group	Fifth decile group	Sixth decile group
Lower boundary of group (£ per week)		120	180	238	304	375
Grossed number of households (thousands)	2,440	2,440	2,430	2,430	2,440	2,430
Total number of households in sample	699	723	724	723	723	691
Total number of persons in sample	902	1,261	1,315	1,572	1,746	1,776
Total number of adults in sample	779	950	1,096	1,206	1,293	1,292
Weighted average number of persons per household	1.3	1.7	1.8	2.2	2.4	2.5
Commodity or service		**Average weekly household expenditure (£)**				
1 Food & non-alcoholic drinks	**21.70**	**27.50**	**32.50**	**36.70**	**40.30**	**42.60**
2 Alcoholic drinks, tobacco & narcotics	**6.40**	**6.00**	**8.80**	**10.30**	**10.50**	**12.00**
3 Clothing & footwear	**6.90**	**8.00**	**10.30**	**15.30**	**18.00**	**21.10**
4 Housing[1], fuel & power	**21.20**	**29.20**	**30.60**	**33.40**	**34.60**	**38.70**
5 Household goods & services	**11.50**	**12.30**	**17.50**	**22.00**	**23.40**	**28.90**
6 Health	**1.70**	**2.40**	**3.50**	**3.50**	**4.40**	**6.10**
7 Transport	**14.50**	**15.80**	**23.50**	**34.90**	**49.20**	**55.80**
8 Communication	**5.70**	**5.80**	**7.70**	**8.50**	**9.30**	**10.40**
9 Recreation & culture	**17.20**	**20.20**	**30.20**	**41.50**	**45.30**	**57.80**
10 Education	**[0.80]**	**1.70**	**1.70**	**0.80**	**2.20**	**2.70**
11 Restaurants & hotels	**11.80**	**11.40**	**15.00**	**21.50**	**27.70**	**33.10**
12 Miscellaneous goods & services	**11.00**	**12.70**	**16.20**	**21.30**	**27.00**	**32.50**
1-12 All expenditure groups	**130.30**	**153.20**	**197.60**	**249.80**	**292.00**	**341.90**
13 Other expenditure items	**14.70**	**15.30**	**23.90**	**37.40**	**42.90**	**56.10**
Total expenditure	**145.00**	**168.50**	**221.50**	**287.20**	**334.90**	**398.00**
Average weekly expenditure per person (£)						
Total expenditure	**112.60**	**99.00**	**123.20**	**133.30**	**140.80**	**157.20**

Note: The commodity and service categories are not comparable with those in publications before 2001-02
1 Excluding mortgage interest payments, council tax and Northern Ireland rates

1.4 Household expenditure by disposable income decile group (cont.) 2002-03

based on weighted data and including children's expenditure

		Seventh decile group	Eighth decile group	Ninth decile group	Highest ten per cent	All house-holds
Lower boundary of group (£ per week)		453	544	665	858	
Grossed number of households (thousands)		2,440	2,430	2,430	2,430	24,350
Total number of households in sample		696	673	650	625	6,927
Total number of persons in sample		2,010	1,967	1,981	2,056	16,586
Total number of adults in sample		1,400	1,434	1,469	1,531	12,450
Weighted average number of persons per household		2.8	2.9	3.0	3.3	2.4
Commodity or service		**Average weekly household expenditure (£)**				
1	**Food & non-alcoholic drinks**	**48.40**	**52.40**	**58.40**	**66.70**	**42.70**
2	**Alcoholic drinks, tobacco & narcotics**	**13.80**	**13.20**	**15.70**	**16.90**	**11.40**
3	**Clothing & footwear**	**25.20**	**32.30**	**35.10**	**50.90**	**22.30**
4	**Housing[1], fuel & power**	**37.30**	**40.60**	**43.10**	**60.10**	**36.90**
5	**Household goods & services**	**34.50**	**38.00**	**45.10**	**69.20**	**30.20**
6	**Health**	**4.40**	**7.10**	**6.00**	**9.10**	**4.80**
7	**Transport**	**69.50**	**82.10**	**106.00**	**141.00**	**59.20**
8	**Communication**	**11.30**	**13.40**	**15.30**	**18.60**	**10.60**
9	**Recreation & culture**	**69.50**	**75.10**	**88.20**	**119.00**	**56.40**
10	**Education**	**3.40**	**4.90**	**7.80**	**25.90**	**5.20**
11	**Restaurants & hotels**	**40.10**	**48.40**	**60.20**	**85.20**	**35.40**
12	**Miscellaneous goods & services**	**36.50**	**45.50**	**53.20**	**74.80**	**33.10**
1-12	**All expenditure groups**	**393.80**	**453.00**	**534.20**	**737.40**	**348.30**
13	**Other expenditure items**	**64.70**	**80.70**	**98.20**	**145.30**	**57.90**
Total expenditure		**458.40**	**533.70**	**632.40**	**882.70**	**406.20**
Average weekly expenditure per person (£)						
Total expenditure		**162.50**	**186.00**	**210.60**	**269.50**	**170.50**

Note: The commodity and service categories are not comparable with those in publications before 2001-02
1 Excluding mortgage interest payments, council tax and Northern Ireland rates

1.5 Household expenditure as a percentage of total expenditure by disposable income decile group 2002-03

based on weighted data and including children's expenditure

	Lowest ten per cent	Second decile group	Third decile group	Fourth decile group	Fifth decile group	Sixth decile group
Lower boundary of group (£ per week)		120	180	238	304	375
Grossed number of households (thousands)	2,440	2,440	2,430	2,430	2,440	2,430
Total number of households in sample	699	723	724	723	723	691
Total number of persons in sample	902	1,261	1,315	1,572	1,746	1,776
Total number of adults in sample	779	950	1,096	1,206	1,293	1,292
Weighted average number of persons per household	1.3	1.7	1.8	2.2	2.4	2.5
Commodity or service			**Percentage of total expenditure**			
1 Food & non-alcoholic drinks	*15*	*16*	*15*	*13*	*12*	*11*
2 Alcoholic drinks, tobacco & narcotics	*4*	*4*	*4*	*4*	*3*	*3*
3 Clothing & footwear	*5*	*5*	*5*	*5*	*5*	*5*
4 Housing[1], fuel & power	*15*	*17*	*14*	*12*	*10*	*10*
5 Household goods & services	*8*	*7*	*8*	*8*	*7*	*7*
6 Health	*1*	*1*	*2*	*1*	*1*	*2*
7 Transport	*10*	*9*	*11*	*12*	*15*	*14*
8 Communication	*4*	*3*	*3*	*3*	*3*	*3*
9 Recreation & culture	*12*	*12*	*14*	*14*	*14*	*15*
10 Education	*[1]*	*1*	*1*	*0*	*1*	*1*
11 Restaurants & hotels	*8*	*7*	*7*	*7*	*8*	*8*
12 Miscellaneous goods & services	*8*	*8*	*7*	*7*	*8*	*8*
1-12 All expenditure groups	*90*	*91*	*89*	*87*	*87*	*86*
13 Other expenditure items	*10*	*9*	*11*	*13*	*13*	*14*
Total expenditure	*100*	*100*	*100*	*100*	*100*	*100*

Note: The commodity and service categories are not comparable with those in publications before 2001-02

1 Excluding mortgage interest payments, council tax and Northern Ireland rates

1.5 Household expenditure as a percentage of total expenditure by disposable income decile group (cont.) — 2002-03

based on weighted data and including children's expenditure

	Seventh decile group	Eighth decile group	Ninth decile group	Highest ten per cent	All house-holds
Lower boundary of group (£ per week)	453	544	665	858	
Grossed number of households (thousands)	2,440	2,430	2,430	2,430	24,350
Total number of households in sample	696	673	650	625	6,927
Total number of persons in sample	2,010	1,967	1,981	2,056	16,586
Total number of adults in sample	1,400	1,434	1,469	1,531	12,450
Weighted average number of persons per household	2.8	2.9	3.0	3.3	2.4
Commodity or service			**Percentage of total expenditure**		
1 **Food & non-alcoholic drinks**	*11*	*10*	*9*	*8*	*11*
2 **Alcoholic drinks, tobacco & narcotics**	*3*	*2*	*2*	*2*	*3*
3 **Clothing & footwear**	*6*	*6*	*6*	*6*	*5*
4 **Housing[1], fuel & power**	*8*	*8*	*7*	*7*	*9*
5 **Household goods & services**	*8*	*7*	*7*	*8*	*7*
6 **Health**	*1*	*1*	*1*	*1*	*1*
7 **Transport**	*15*	*15*	*17*	*16*	*15*
8 **Communication**	*2*	*3*	*2*	*2*	*3*
9 **Recreation & culture**	*15*	*14*	*14*	*13*	*14*
10 **Education**	*1*	*1*	*1*	*3*	*1*
11 **Restaurants & hotels**	*9*	*9*	*10*	*10*	*9*
12 **Miscellaneous goods & services**	*8*	*9*	*8*	*8*	*8*
1-12 **All expenditure groups**	*86*	*85*	*84*	*84*	*86*
13 **Other expenditure items**	*14*	*15*	*16*	*16*	*14*
Total expenditure	*100*	*100*	*100*	*100*	*100*

Note: The commodity and service categories are not comparable with those in publications before 2001-02
1 Excluding mortgage interest payments, council tax and Northern Ireland rates

Chapter 2

Expenditure by age and income

Income is not adjusted to take into account the different composition of households (equivalisation) as done in some other income analyses

- Average weekly expenditure for all households in 2002-03 was £406. This varied by age from the highest at £497 in households where the reference person was aged 30 to 49 to the lowest at £177 in households where the reference person was aged 75 or over.

- The only categories where households with a reference person aged under 30 spent more than other age groups were **housing, fuel and power** (excluding mortgage interest payments, council tax/rates) **and communications.**

- The proportion of spending going on **food and non-alcoholic drinks** rose with age from eight per cent where the reference person was aged under 30 to 17 per cent for those aged 75 or over. The proportion spent on **restaurants and hotels** decreased with age from 11 per cent of total spending where the reference person was aged under 30 to six per cent for those aged 75 and over.

- Households where the reference person was aged 50 to 64 spent more than other age groups on **fresh fruit, vegetables and potatoes.** Those in the under 30 age group spent the least. Spending on **chocolate and confectionery** was highest for the 30 to 49 age group at £2.30 a week and lowest for those aged 75 or over at £1.20 a week.

- **Transport** was the largest item of expenditure for households where the reference person was aged under 65. Households where the reference person was aged 50 to 64 spent more than other groups on purchases of **new cars/ vans,** but for **second-hand vehicles** the highest spending group were households with a reference person aged 30 to 49.

- Spending on **newspapers** was highest for households with a reference person aged 65 to 74 at nearly £3 a week and lowest for the under 30 age group at 70p a week.

2.1 Expenditure by age of HRP

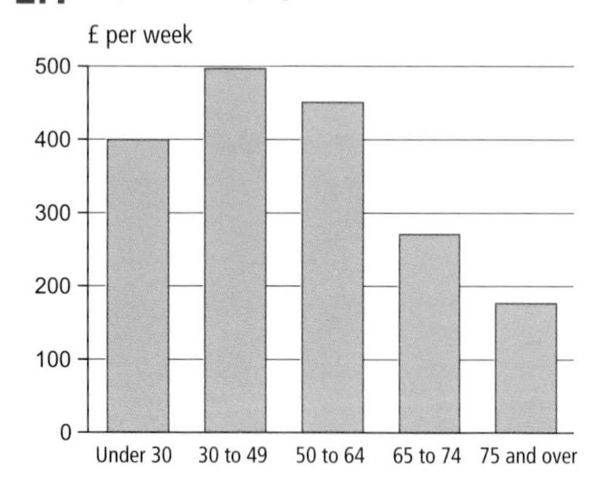

2.2 Food and transport as a percentage of total expenditure by age of HRP

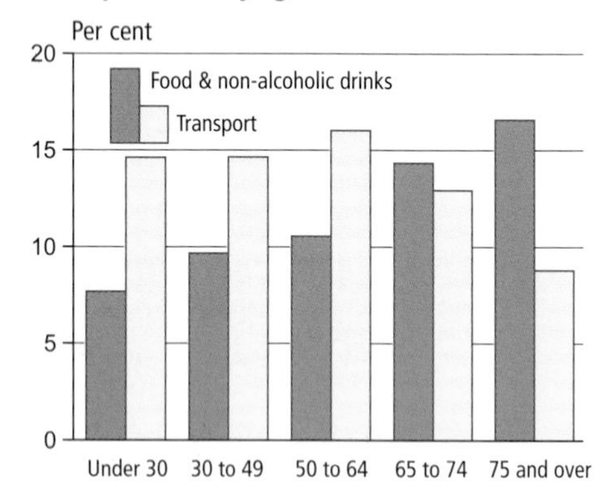

2.3 Recreation & culture and alcohol & tobacco as a percentage of total expenditure by age of HRP

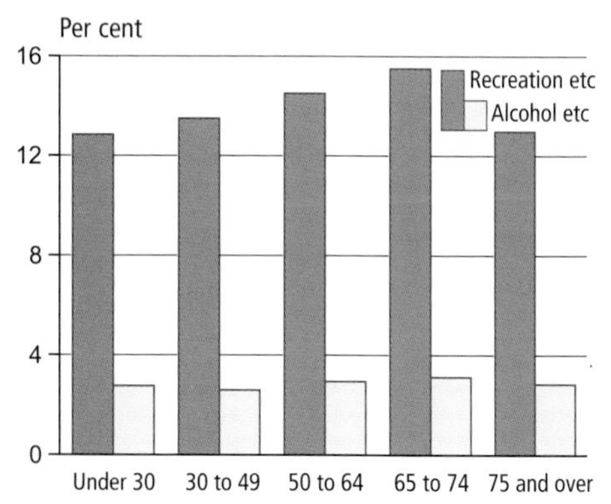

Chapter 2
Expenditure by age and income

This chapter analyses household expenditure by age of household reference person. In addition, **Tables 2.4 to 2.8** contain an analysis of average weekly expenditure for each group by income.

From 2001-02, the concept of household reference person (HRP) was adopted on all government-sponsored surveys, in place of head of household. For a definition of HRP see Appendix D Definitions, page 173.

Characteristics of households

Table 2.1 shows how the number of households varied across the age groups. Ten per cent had a HRP aged under 30. Households with a HRP aged 30 to 49 were the largest group, forming 39 per cent of the total, 25 per cent had a HRP aged 50 to 64, and a further 26 per cent had a HRP aged 65 or over. The average number of people per household was highest for households with a HRP aged 30 to 49, 3 people, falling to 1.4 for those households with a HRP aged 75 or over. The average for all households was 2.4 people per household.

Aggregate expenditure patterns

Figure 2.1 shows that average weekly household expenditure was lowest at £177 a week for households where the HRP was aged 75 or over and highest at £497 a week where the HRP was 30 to 49. **Table 2.1** shows that expenditure on transport was the largest item of spending for households with a HRP aged under 65. Households with a HRP aged between 65 and 74 spent most on recreation and culture while the largest item of expenditure for the oldest age group was food and non-alcoholic drinks.

Table 2.2 shows expenditure on the main commodities as a percentage of total expenditure by age of HRP. **Figure 2.2** shows the proportions for two items, transport and food and non-alcoholic drinks. As a proportion of total expenditure, spending on transport was fairly constant for age groups under 65, at around 15 per cent of total expenditure. However, it fell to nine per cent of total expenditure for households with a HRP aged 75 or over. In contrast, spending on food and non-alcoholic drinks increased steadily with age of HRP, from eight per cent for the under 30 age group to 17 per cent for those aged 75 or over.

Figure 2.3 shows the proportion spent on two further items, recreation and culture and alcoholic drinks and tobacco, by age of HRP. The proportion spent on recreation and culture rose steadily through the age groups, from 13 per cent in the under 30 age group to 16 per cent for those aged 65 to 74. It then fell back to 13 per cent for the oldest age group, households with an HRP aged 75 or over. However, spending on alcohol was unique among the main categories in that the proportion spent was independent of the age of HRP, all age groups spending three per cent of their total on this item.

2.4 Sports admissions and pets by age of HRP

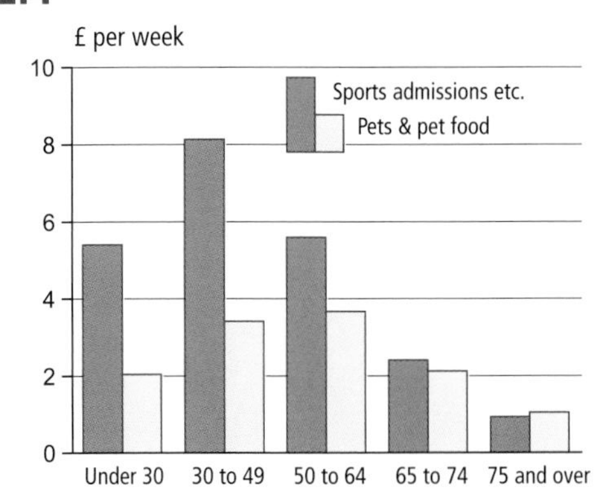

Detailed expenditure patterns

Sports admissions and pets and pet food

Table 2.3 gives a detailed breakdown of expenditure for each age group. **Figure 2.4** compares expenditure on sports admissions, subscriptions and leisure class fees with expenditure on pets and pet food across all age groups. It shows that for all households with a HRP under 50, spending on sports admissions was on average over double that on pets and pet food, £7.50 a week compared to £3.10. However, for the older age groups, spending on these items was much more even, and indeed for the oldest age group, spending on pets and pet food was greater at £1.00 a week than that on sports admissions, at 90p a week.

2.5 Gambling payments and books & diaries by age of HRP

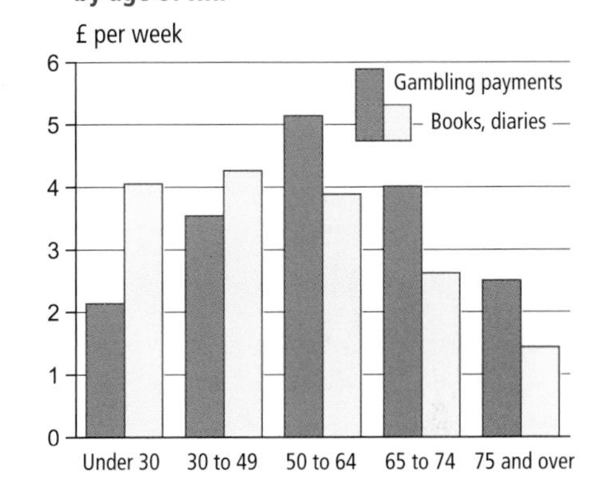

Gambling payments and books and diaries

Figure 2.5 compares spending on gambling payments with books, diaries, address books, cards etc. The largest spenders on gambling payments were households with a HRP aged 50 to 64, at £5.10 a week, almost 30 per cent more than the next highest spenders, those aged 65 to 74, at £4.00 a week. However, the pattern was different for expenditure on books and diaries. Expenditure was highest on this item for households with a HRP in the 30 to 49 age group, that spent most at £4.30 a week. Spending decreased after this to £1.40 for the 75 or over age group.

Men's and women's outer garments

Figure 2.6 shows spending on men's and women's outer garments. Expenditure on women's outer garments was higher than on men's for all age groups. However, the age group for which spending was highest was different for the two items. Spending on women's outer garments was highest in the 50 to 64 age group at £10.20 a week, and for men's, most was spent by the 30 to 49 age group at £5.80 a week. Spending on men's outer garments was 66 per cent that of women's for households with a HRP aged under 30 and fell to 45 per cent for the 50 to 64 age group.

2.6 Men's and women's outer garments by age of HRP

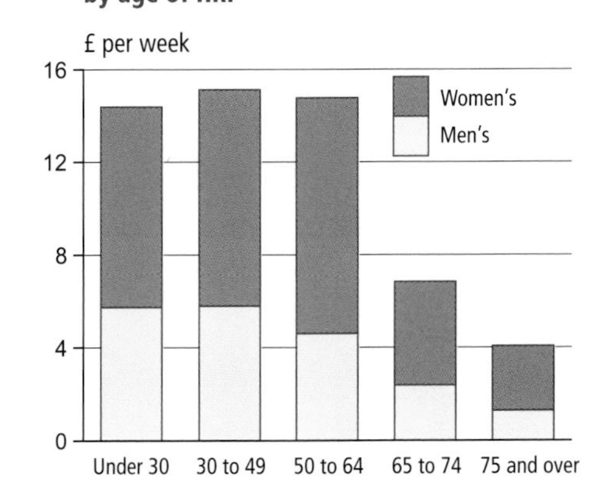

2.7a Total expenditure by age of HRP: lowest income quintile group

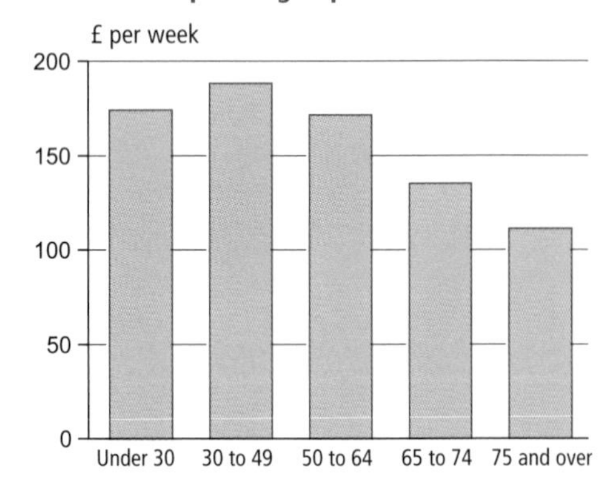

2.7b Housing, fuel & power and food by age of HRP: lowest income quintile group

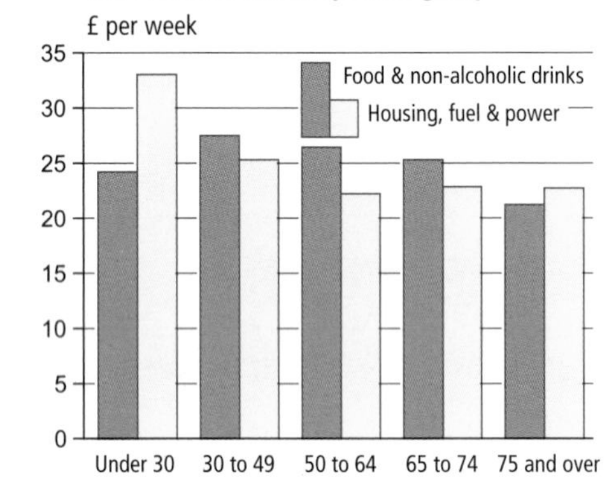

2.7c Recreation & culture and transport by age of HRP: lowest income quintile group

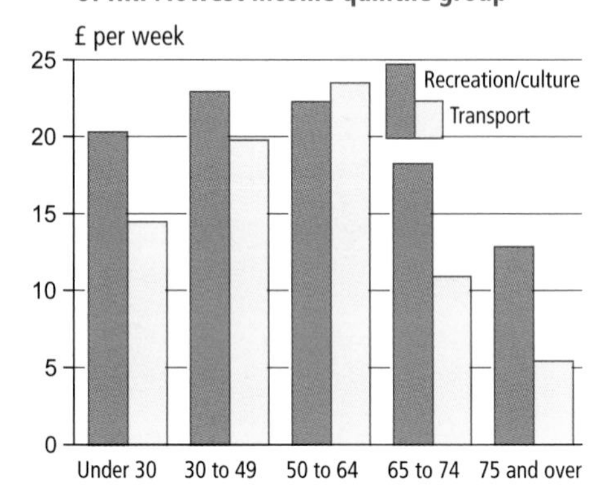

Expenditure by age and income

Much of the variation in spending with age is the result of differences in income. **Tables 2.4 to 2.8** contain an analysis of expenditure by gross income quintile group for each age group. The data are averaged over the last two years to increase the sample size.

Low income households

Figure 2.7a shows how total expenditure varied with age for the fifth of all households with the lowest incomes. It shows that households with a HRP aged 65 or over spent £122 a week on average, markedly less than those with similar income but a younger HRP, who spent £176 a week on average. The highest spending individual age group for households in the lowest income quintile was the 30 to 49 age group, who spent £189 a week.

Figures 2.7b and c illustrate expenditure on selected items for the lowest income quintile group, by age of HRP. Spending on food and non-alcoholic drinks was similar across all age groups, peaking at £27.60 a week for the 30 to 49 age group and falling to £21.30 for households with a HRP aged 75 or over. In contrast, housing, fuel and power expenditure was much higher for households with a HRP aged under 30 than for any other age group, at £33.20 a week. Spending on this item fell to £25.40 a week for the 30 to 49 age group, and averaged £22.80 for all households with a HRP older than this.

Tables 2.4 to 2.8 show that for households in the lowest income group, housing, fuel and power was the largest item of expenditure in both the youngest and oldest age groups, at £33.20 and £22.80 a week respectively. Both of these figures are around 20 per cent of the total spending for their respective age groups. For the age groups in between, households spent most on food and non-alcoholic drinks. For the 30 to 49 age group, the spend on food was £27.60 a week, and this represented just under 15 per cent of the total spend. However, the spend of £25.40 on this item by households in the 65 to 74 age group was nearly 19 per cent of their total expenditure.

Figure 2.7c compares expenditure on recreation and culture and transport by age of HRP for the lowest income group. Transport expenditure was highest in the 50 to 64 age group at £23.60 a week. This was the only age group for which transport spending exceeded spending on recreation and culture. Recreation and culture expenditure peaked in the 30 to 49 age group and decreased thereafter much less steeply than transport spending, which fell to £5.50 for households with a HRP aged 75 or over.

Middle income households

Figure 2.8a shows total expenditure and **Figures 2.8b and 2.8c** show expenditure on selected items for the third (middle) income quintile group by age of HRP. Spending was more constant for households with a HRP below the age of 65, at £373 a week, when compared to the same age groups in the lowest income band. For older age groups, the spending again decreased, but not as markedly as for the lowest income group. Households with a HRP aged 75 or over spent the least of all ages, at £262 a week, 30 per cent less than the highest spenders, the 30 to 49 age group.

Spending on housing, fuel and power was highest for the youngest age group, at £57 a week, and decreased more noticeably for the older age groups than was the case for the lowest income households. Expenditure on food and non-alcoholic drinks was highest for the 65 to 74 age group, although spending was fairly constant for all households with a HRP aged over 30.

Expenditure on both transport and recreation and culture was more uniform across age groups in middle income households than in lowest income households. For both items, the highest spenders were those households with a HRP aged 50 to 64 (£58.70 a week for transport and £55.70 a week for recreation and culture). Expenditure on both items was similar for most age groups, except for the under 30s, where spending on recreation and culture was £41.80, 81 per cent of the amount spent on transport by the same group. The 65 to 74 age group was the only one in which spending on transport exceeded that on recreation and culture.

2.8a Total expenditure by age of HRP: third income quintile group

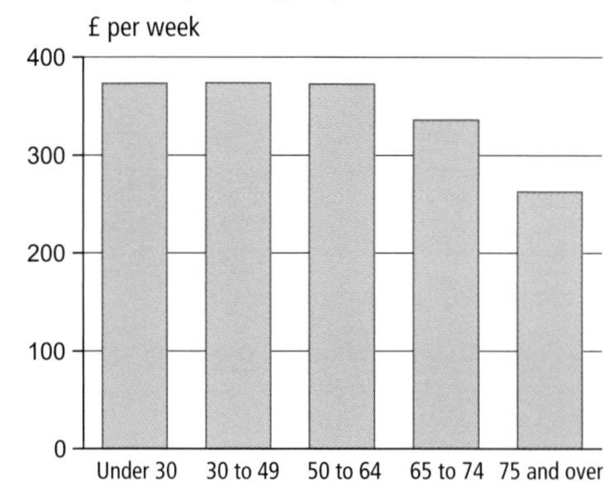

2.8b Housing, fuel & power and food by age of HRP: third income quintile group

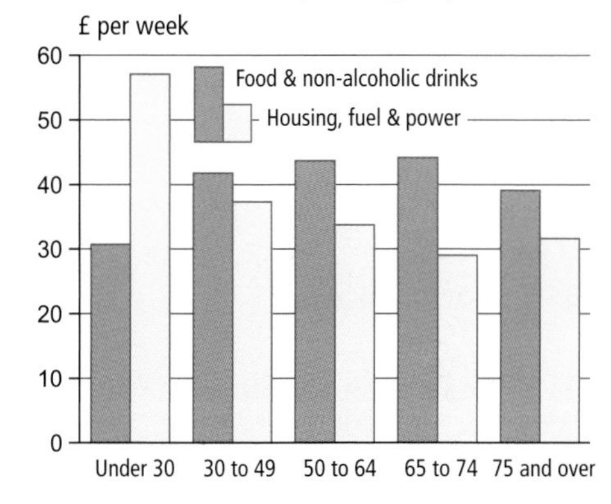

2.8c Recreation & culture and transport by age of HRP: third income quintile group

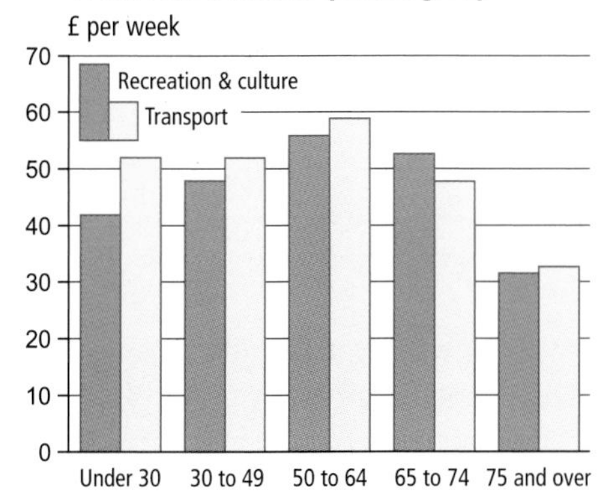

2.1 Household expenditure by age of household reference person 2002-03

based on weighted data and including children's expenditure

	Under 30	30 and under 50	50 and under 65	65 and under 75	75 or over	All households
Grossed number of households (thousands)	2,550	9,520	6,040	3,280	2,960	24,350
Total number of households in sample	680	2,752	1,682	1,006	807	6,927
Total number of persons in sample	1,631	8,417	3,649	1,721	1,168	16,586
Total number of adults in sample	1,163	5,096	3,331	1,692	1,168	12,450
Weighted average number of persons per household	2.4	3.0	2.2	1.7	1.4	2.4
Commodity or service	**Average weekly household expenditure (£)**					
1 Food & non-alcoholic drinks	**30.90**	**48.20**	**47.80**	**38.90**	**29.40**	**42.70**
2 Alcoholic drinks, tobacco & narcotics	**11.20**	**13.00**	**13.40**	**8.50**	**5.10**	**11.40**
3 Clothing & footwear	**23.00**	**29.20**	**24.00**	**11.60**	**7.80**	**22.30**
4 Housing[1], fuel & power	**55.80**	**38.90**	**35.30**	**29.30**	**25.90**	**36.90**
5 Household goods & services	**22.70**	**35.60**	**36.90**	**22.00**	**14.90**	**30.20**
6 Health	**2.40**	**4.60**	**6.30**	**5.90**	**3.50**	**4.80**
7 Transport	**58.60**	**72.90**	**72.40**	**35.10**	**15.60**	**59.20**
8 Communication	**13.30**	**12.80**	**10.60**	**6.90**	**5.10**	**10.60**
9 Recreation & culture	**51.40**	**67.10**	**65.60**	**42.10**	**23.00**	**56.40**
10 Education	**4.80**	**7.40**	**6.40**	**[1.00]**	**..**	**5.20**
11 Restaurants & hotels	**43.80**	**44.00**	**38.50**	**20.10**	**11.50**	**35.40**
12 Miscellaneous goods & services	**31.80**	**41.20**	**34.70**	**22.20**	**16.90**	**33.10**
1-12 All expenditure groups	**349.80**	**414.90**	**392.00**	**243.50**	**159.40**	**348.30**
13 Other expenditure items	**50.20**	**82.10**	**59.30**	**27.40**	**17.80**	**57.90**
Total expenditure	**400.10**	**496.90**	**451.40**	**270.90**	**177.20**	**406.20**
Average weekly expenditure per person (£)						
Total expenditure	**168.90**	**165.60**	**202.50**	**157.30**	**122.60**	**170.50**

Note: The commodity and service categories are not comparable with those in publications before 2001-02
1 Excluding mortgage interest payments, council tax and Northern Ireland rates

2.2 Household expenditure as a percentage of total expenditure by age of household reference person

2002-03

based on weighted data and including children's expenditure

	Under 30	30 and under 50	50 and under 65	65 and under 75	75 or over	All households
Grossed number of households (thousands)	2,548	9,524	6,037	3,276	2,961	24,346
Total number of households in sample	680	2,752	1,682	1,006	807	6,927
Total number of persons in sample	1,631	8,417	3,649	1,721	1,168	16,586
Total number of adults in sample	1,163	5,096	3,331	1,692	1,168	12,450
Weighted average number of persons per household	2.4	3.0	2.2	1.7	1.4	2.4
Commodity or service			**Percentage of total expenditure**			
1 Food & non-alcoholic drinks	*8*	*10*	*11*	*14*	*17*	*11*
2 Alcoholic drinks, tobacco & narcotics	*3*	*3*	*3*	*3*	*3*	*3*
3 Clothing & footwear	*6*	*6*	*5*	*4*	*4*	*5*
4 Housing, fuel & power	*14*	*8*	*8*	*11*	*15*	*9*
5 Household goods & services	*6*	*7*	*8*	*8*	*8*	*7*
6 Health	*1*	*1*	*1*	*2*	*2*	*1*
7 Transport	*15*	*15*	*16*	*13*	*9*	*15*
8 Communication	*3*	*3*	*2*	*3*	*3*	*3*
9 Recreation & culture	*13*	*14*	*15*	*16*	*13*	*14*
10 Education	*1*	*1*	*1*	*[0]*	..	*1*
11 Restaurants & hotels	*11*	*9*	*9*	*7*	*6*	*9*
12 Miscellaneous goods & services	*8*	*8*	*8*	*8*	*10*	*8*
1-12 All expenditure groups	*87*	*83*	*87*	*90*	*90*	*86*
13 Other expenditure items	*13*	*17*	*13*	*10*	*10*	*14*
Total expenditure	*100*	*100*	*100*	*100*	*100*	*100*

Note: The commodity and service categories are not comparable with those in publications before 2001-02
1 Excluding mortgage interest payments, council tax and Northern Ireland rates

2.3 Detailed household expenditure by age of household reference person 2002-03

based on weighted data and including children's expenditure

	Under 30	30 and under 50	50 and under 65	65 and under 75	75 or over	All house-holds
Grossed number of households (thousands)	2,550	9,520	6,040	3,280	2,960	24,350
Total number of households in sample	680	2,752	1,682	1,006	807	6,927
Total number of persons in sample	1,631	8,417	3,649	1,721	1,168	16,586
Total number of adults in sample	1,163	5,096	3,331	1,692	1,168	12,450
Weighted average number of persons per household	2.4	3.0	2.2	1.7	1.4	2.4
Commodity or service	**Average weekly household expenditure (£)**					
1 Food & non-alcoholic drinks	**30.90**	**48.20**	**47.80**	**38.90**	**29.40**	**42.70**
1.1 Food	27.70	43.80	43.90	36.00	27.30	39.10
1.1.1 Bread, rice and cereals	3.20	4.60	4.00	3.20	2.30	3.90
1.1.2 Pasta products	0.50	0.50	0.30	0.10	0.10	0.30
1.1.3 Buns, cakes, biscuits etc.	1.60	3.20	3.00	2.70	2.40	2.80
1.1.4 Pastry (savoury)	0.70	0.90	0.60	0.30	0.20	0.60
1.1.5 Beef (fresh, chilled or frozen)	0.80	1.30	1.80	1.50	0.90	1.40
1.1.6 Pork (fresh, chilled or frozen)	0.30	0.60	0.80	0.70	0.40	0.60
1.1.7 Lamb (fresh, chilled or frozen)	0.20	0.50	0.80	0.70	0.70	0.60
1.1.8 Poultry (fresh, chilled or frozen)	1.00	1.70	1.80	1.20	0.80	1.50
1.1.9 Bacon and ham	0.50	0.80	1.10	1.00	0.70	0.90
1.1.10 Other meat and meat preparations	3.50	5.60	5.30	4.00	3.30	4.80
1.1.11 Fish and fish products	1.10	1.70	2.10	2.10	1.70	1.80
1.1.12 Milk	1.50	2.40	2.20	2.00	1.90	2.10
1.1.13 Cheese and curd	1.10	1.60	1.60	1.20	0.80	1.40
1.1.14 Eggs	0.30	0.40	0.50	0.40	0.30	0.40
1.1.15 Other milk products	1.00	1.70	1.50	1.30	1.00	1.40
1.1.16 Butter	0.10	0.20	0.30	0.40	0.30	0.30
1.1.17 Margarine and other vegetable fats	0.20	0.40	0.50	0.50	0.40	0.40
1.1.18 Peanut butter	0.00	0.00	0.00	[0.00]	[0.00]	0.00
1.1.19 Cooking oils and fats	0.10	0.20	0.20	0.20	0.10	0.20
1.1.20 Fresh fruit	1.30	2.40	3.00	2.70	2.10	2.40
1.1.21 Other fresh, chilled or frozen fruits	0.10	0.20	0.30	0.20	0.10	0.20
1.1.22 Dried fruit and nuts	0.10	0.30	0.40	0.40	0.40	0.30
1.1.23 Preserved fruit and fruit based products	0.10	0.10	0.20	0.20	0.20	0.10
1.1.24 Fresh vegetables	2.00	3.30	3.70	2.80	2.00	3.00
1.1.25 Dried vegetables and other preserved or processed vegetables	0.90	1.30	1.10	0.80	0.50	1.00
1.1.26 Potatoes	0.50	0.80	0.90	0.80	0.60	0.70
1.1.27 Other tubers and products of tuber vegetables	1.10	1.60	1.10	0.70	0.40	1.20
1.1.28 Sugar and sugar products	0.10	0.30	0.30	0.30	0.30	0.30
1.1.29 Jams, marmalades	0.10	0.20	0.30	0.30	0.20	0.20
1.1.30 Chocolate	0.90	1.60	1.20	1.10	0.80	1.20
1.1.31 Confectionery products	0.50	0.70	0.60	0.50	0.40	0.60
1.1.32 Edible ices and ice cream	0.30	0.60	0.50	0.40	0.30	0.50
1.1.33 Other food products	1.80	2.20	1.90	1.30	0.90	1.80
1.2 Non-alcoholic drinks	3.20	4.40	3.90	2.90	2.00	3.70
1.2.1 Coffee	0.30	0.50	0.70	0.50	0.30	0.50
1.2.2 Tea	0.20	0.40	0.60	0.60	0.50	0.50
1.2.3 Cocoa and powdered chocolate	0.10	0.10	0.10	0.10	0.10	0.10
1.2.4 Fruit and vegetable juices, mineral waters	1.10	1.40	1.20	0.80	0.60	1.20
1.2.5 Soft drinks	1.50	2.00	1.40	0.90	0.50	1.50

Note: The commodity and service categories are not comparable with those in publications before 2001-02
The numbering system is sequential, it does not use actual COICOP codes

2.3 Detailed household expenditure by age of household reference person (cont.) 2002-03

based on weighted data and including children's expenditure

		Under 30	30 and under 50	50 and under 65	65 and under 75	75 or over	All house-holds
	Commodity or service	Average weekly household expenditure (£)					
2	**Alcoholic drink, tobacco & narcotics**	**11.20**	**13.00**	**13.40**	**8.50**	**5.10**	**11.40**
2.1	Alcoholic drinks	4.90	6.70	7.10	4.60	3.30	5.90
2.1.1	Spirits and liqueurs (brought home)	0.70	1.00	1.50	1.60	1.30	1.20
2.1.2	Wines, fortified wines (brought home)	1.90	3.30	3.80	2.00	1.50	2.90
2.1.3	Beer, lager, ciders and Perry (brought home)	2.10	2.20	1.70	1.00	0.60	1.70
2.1.4	Alcopops (brought home)	0.20	0.20	0.10	[0.10]	..	0.10
2.2	Tobacco and narcotics	6.20	6.30	6.30	3.90	1.80	5.40
2.2.1	Cigarettes	5.70	5.70	5.30	3.30	1.50	4.80
2.2.2	Cigars, other tobacco products and narcotics	0.60	0.60	1.00	0.70	0.20	0.60
3	**Clothing & footwear**	**23.00**	**29.20**	**24.00**	**11.60**	**7.80**	**22.30**
3.1	Clothing	18.90	23.80	20.10	9.70	6.00	18.30
3.1.1	Men's outer garments	5.70	5.80	4.60	2.40	1.30	4.50
3.1.2	Men's under garments	0.30	0.40	0.50	0.40	0.40	0.40
3.1.3	Women's outer garments	8.60	9.30	10.20	4.40	2.80	8.00
3.1.4	Women's under garments	1.00	1.50	1.50	1.00	0.80	1.30
3.1.5	Boys' outer garments (5-15)	0.30	1.70	0.40	0.20	..	0.80
3.1.6	Girls' outer garments (5-15)	0.40	2.20	0.80	0.20	..	1.10
3.1.7	Infants' outer garments (under 5)	1.00	0.90	0.40	0.20	..	0.60
3.1.8	Children's under garments (under 16)	0.30	0.70	0.30	0.10	..	0.40
3.1.9	Accessories	0.80	0.90	0.70	0.30	0.20	0.70
3.1.10	Haberdashery, clothing materials and clothing hire	[0.10]	0.20	0.30	0.20	0.10	0.20
3.1.11	Dry cleaners, laundry and dyeing	0.20	0.30	0.40	0.30	0.20	0.30
3.2	Footwear	4.10	5.40	3.90	1.90	1.80	4.00
4	**Housing (net)[1], fuel & power**	**55.80**	**38.90**	**35.30**	**29.30**	**25.90**	**36.90**
4.1	Actual rentals for housing	52.60	21.10	12.50	17.10	20.00	21.60
4.1.1	Gross rent	52.60	21.10	12.50	17.10	20.00	21.60
4.1.2	*less* housing benefit, rebates & allowances rec'd	14.30	8.00	6.60	10.20	14.30	9.40
4.1.3	Net rent	38.30	13.10	5.80	6.90	5.70	12.20
4.1.4	Second dwelling rent	..	..	..	..	..	..
4.2	Maintenance and repair of dwelling	3.40	8.20	10.50	6.10	4.80	7.50
4.3	Water supply and miscellaneous services relating to the dwelling	5.10	5.20	5.90	5.30	5.40	5.40
4.4	Electricity, gas and other fuels	9.10	12.40	12.90	11.00	10.00	11.70
4.4.1	Electricity	4.50	6.20	6.20	5.70	5.00	5.80
4.4.2	Gas	4.20	5.50	5.70	4.70	4.30	5.10
4.4.3	Other fuels	0.40	0.70	1.10	0.70	0.70	0.80

Note: The commodity and service categories are not comparable with those in publications before 2001-02
The numbering system is sequential, it does not use actual COICOP codes
1 Excluding mortgage interest payments, council tax and Northern Ireland rates

2.3 Detailed household expenditure by age of household reference person (cont.) 2002-03

based on weighted data and including children's expenditure

Commodity or service		Under 30	30 and under 50	50 and under 65	65 and under 75	75 or over	All house-holds
		Average weekly household expenditure (£)					
5	**Household goods & services**	**22.70**	**35.60**	**36.90**	**22.00**	**14.90**	**30.20**
5.1	Furniture and furnishings, carpets and other floor coverings	12.90	20.50	18.80	9.30	6.40	16.10
5.1.1	Furniture and furnishings	10.50	16.50	14.20	6.80	4.40	12.50
5.1.2	Floor coverings	2.30	4.00	4.60	2.50	2.00	3.50
5.1.3	Repair of furniture, furnishings and floor coverings	..	..	..	..	..	..
5.2	Household textiles	1.30	2.50	2.50	1.70	1.00	2.10
5.3	Household appliances	3.40	3.10	4.10	3.10	1.80	3.20
5.4	Glassware, tableware and household utensils	1.20	1.80	2.40	1.20	0.60	1.60
5.5	Tools and equipment for house and garden	1.60	2.70	3.90	2.60	1.00	2.70
5.6	Goods and services for routine household maintenance	2.40	5.10	5.20	4.10	4.10	4.60
5.6.1	Cleaning materials	1.40	2.40	2.20	1.80	1.20	2.00
5.6.2	Household goods and hardware	0.70	1.20	1.20	1.00	0.60	1.10
5.6.3	Domestic services, carpet cleaning	0.30	1.50	1.70	1.20	2.30	1.50
6	**Health**	**2.40**	**4.60**	**6.30**	**5.90**	**3.50**	**4.80**
6.1	Medical products, appliances and equipment	1.90	2.90	4.30	3.60	2.30	3.20
6.1.1	Medicines, prescriptions and healthcare products	1.10	1.50	2.00	1.20	1.10	1.50
6.1.2	Spectacles, lenses, accessories and repairs	0.80	1.40	2.10	2.20	[1.10]	1.60
6.1.3	Non-optical appliances and equipment (e.g. wheelchairs, batteries for hearing aids, etc.)	..	..	[0.20]	..	..	0.10
6.2	Hospital services	0.50	1.70	2.00	2.30	1.20	1.70
7	**Transport**	**58.60**	**72.90**	**72.40**	**35.10**	**15.60**	**59.20**
7.1	Purchase of vehicles	24.40	33.50	33.70	14.70	5.10	26.60
7.1.1	Purchase of new cars and vans	6.50	13.10	16.00	8.30	3.40	11.30
7.1.2	Purchase of second hand cars or vans	17.20	19.10	16.90	6.00	1.80	14.50
7.1.3	Purchase of motorcycles and other vehicles	[0.70]	1.30	0.80	..	..	0.90
7.2	Operation of personal transport	21.40	29.00	29.60	16.40	7.70	24.10
7.2.1	Spares and accessories	1.70	2.40	2.90	1.70	0.40	2.10
7.2.2	Petrol, diesel and other motor oils	14.30	18.20	18.10	9.30	3.90	14.80
7.2.3	Repairs and servicing	4.20	5.90	6.30	4.10	2.60	5.20
7.2.4	Other motoring costs	1.20	2.40	2.30	1.30	0.80	1.90
7.3	Transport services	12.70	10.40	9.10	4.00	2.80	8.50
7.3.1	Rail and tube fares	3.50	2.30	1.80	0.60	0.20	1.80
7.3.2	Bus and coach fares	2.20	1.80	1.40	0.80	0.50	1.40
7.3.4	Combined fares	2.20	0.90	0.60	..	..	0.80
7.3.5	Other travel and transport	4.80	5.40	5.20	2.30	2.10	4.50

Note: The commodity and service categories are not comparable with those in publications before 2001-02
The numbering system is sequential, it does not use actual COICOP codes

2.3 Detailed household expenditure by age of household reference person (cont.) 2002-03

based on weighted data and including children's expenditure

		Under 30	30 and under 50	50 and under 65	65 and under 75	75 or over	All households
Commodity or service		Average weekly household expenditure (£)					
8	**Communication**	**13.30**	**12.80**	**10.60**	**6.90**	**5.10**	**10.60**
8.1	Postal services	0.30	0.40	0.60	0.60	0.50	0.50
8.2	Telephone and telefax equipment	0.60	1.00	0.50	0.30	..	0.60
8.3	Telephone and telefax services	12.50	11.50	9.50	5.90	4.50	9.50
9	**Recreation & culture**	**51.40**	**67.10**	**65.60**	**42.10**	**23.00**	**56.40**
9.1	Audio-visual, photographic and information processing equipment	13.00	10.60	8.00	2.80	2.50	8.20
9.1.1	Audio equipment and accessories, CD players	2.90	3.00	2.50	0.80	0.20	2.30
9.1.2	TV, video and computers	9.20	6.30	4.40	1.90	2.10	5.00
9.1.3	Photographic, cinematographic & optical equip't	0.80	1.30	1.00	0.20	[0.10]	0.90
9.2	Other major durables for recreation and culture	[0.20]	2.00	3.10	..	..	1.80
9.3	Other recreational items and equipment, gardens and pets	7.40	12.50	12.40	6.00	3.30	10.00
9.3.1	Games, toys and hobbies	1.90	3.30	2.40	0.90	0.30	2.20
9.3.2	Computer software and games	1.50	1.60	1.00	[0.20]	..	1.10
9.3.3	Equipment for sport, camping and open-air recreation	0.70	1.30	0.80	0.20	..	0.80
9.3.4	Horticultural goods, garden equipment and plants	1.20	3.00	4.50	2.70	1.80	3.00
9.3.5	Pets and pet food	2.00	3.40	3.70	2.10	1.00	2.90
9.4	Recreational and cultural services	16.60	21.50	19.30	12.20	5.50	17.20
9.4.1	Sports admissions, subscriptions and leisure class fees	5.40	8.10	5.60	2.40	0.90	5.60
9.4.2	Cinema, theatre and museums etc.	1.80	2.00	1.80	1.20	0.60	1.60
9.4.3	TV, video, satellite rental, cable subscriptions, TV licences and the Internet	5.40	6.00	5.30	3.70	1.10	4.80
9.4.4	Miscellaneous entertainments	1.30	1.20	1.10	0.70	0.40	1.00
9.4.5	Development of film, deposit for film development, passport photos, holiday and school photos	0.70	0.60	0.40	0.20	[0.10]	0.50
9.4.6	Gambling payments	2.10	3.50	5.10	4.00	2.50	3.70
9.5	Newspapers, books and stationery	5.80	7.10	7.30	6.30	4.40	6.60
9.5.1	Books, diaries, address books, cards etc.	4.00	4.30	3.90	2.60	1.40	3.60
9.5.2	Newspapers	0.70	1.50	2.30	2.90	2.30	1.90
9.5.3	Magazines and periodicals	1.00	1.30	1.20	0.80	0.70	1.10
9.6	Package holidays	8.40	13.50	15.50	13.20	7.20	12.70
9.6.1	Package holidays - UK	..	0.60	0.90	1.50	1.80	0.90
9.6.2	Package holidays - abroad	8.30	12.80	14.60	11.70	5.30	11.70
10	**Education**	**4.80**	**7.40**	**6.40**	**[1.00]**	**..**	**5.20**
10.1	Education fees	4.60	6.80	6.20	[1.00]	..	4.90
10.2	Payments for school trips, other ad-hoc expenditure	[0.20]	0.60	[0.10]	..	..	0.30

Note: The commodity and service categories are not comparable with those in publications before 2001-02
The numbering system is sequential, it does not use actual COICOP codes

2.3 Detailed household expenditure by age of household reference person (cont.) 2002-03

based on weighted data and including children's expenditure

	Under 30	30 and under 50	50 and under 65	65 and under 75	75 or over	All house-holds
Commodity or service			Average weekly household expenditure (£)			
11 Restaurants & hotels	**43.80**	**44.00**	**38.50**	**20.10**	**11.50**	**35.40**
11.1 Catering services	40.80	38.00	32.80	15.70	8.80	30.50
11.1.1 Restaurant and café meals	11.70	12.80	13.10	8.40	5.40	11.30
11.1.2 Alcoholic drinks (away from home)	14.50	10.10	10.10	4.90	2.00	8.90
11.1.3 Take away meals eaten at home	5.60	5.00	3.00	1.20	1.00	3.60
11.1.4 Other take-away and snack food	6.30	6.70	3.10	1.00	0.30	4.20
11.1.5 Contract catering (food)	[0.30]	0.50	..	..	..	0.70
11.1.6 Canteens	2.40	3.00	1.50	0.20	[0.10]	1.80
11.2 Accommodation services	3.00	5.90	5.80	4.40	2.70	5.00
11.2.1 Holiday in the UK	1.10	2.70	2.90	2.90	1.90	2.50
11.2.2 Holiday abroad	1.90	3.20	2.80	1.40	[0.80]	2.40
11.2.3 Room hire	..	[0.10]	..	..	..	0.00
12 Miscellaneous goods & services	**31.80**	**41.20**	**34.70**	**22.20**	**16.90**	**33.10**
12.1 Personal care	8.70	10.50	9.00	6.40	4.80	8.70
12.1.1 Hairdressing, beauty treatment	2.20	2.70	3.00	2.60	2.50	2.70
12.1.2 Toilet paper	0.60	0.80	0.70	0.60	0.40	0.70
12.1.3 Toiletries and soap	1.80	2.40	2.00	1.30	0.90	1.90
12.1.4 Baby toiletries and accessories (disposable)	1.00	0.90	0.20	0.10	0.00	0.50
12.1.5 Hair products, cosmetics and related electrical appliances	3.20	3.60	3.10	1.80	1.00	2.90
12.2 Personal effects	3.50	3.10	3.70	1.60	0.90	2.80
12.3 Social protection	2.50	4.60	0.80	1.00	1.90	2.60
12.4 Insurance	13.30	16.80	16.90	10.90	8.70	14.70
12.4.1 Household insurances - structural, contents and appliances	2.80	5.20	5.10	3.90	3.20	4.50
12.4.2 Medical insurance premiums	0.30	1.00	2.00	1.60	2.30	1.40
12.4.3 Vehicle insurance including boat insurance	10.00	10.50	9.50	4.90	3.20	8.60
12.4.4 Non-package holiday, other travel insurance	..	0.20	0.30	[0.50]	..	0.20
12.5 Other services	3.70	6.10	4.30	2.30	0.60	4.30
12.5.1 Moving house	2.40	3.60	2.20	1.40	[0.30]	2.40
12.5.2 Bank, building society, post office, credit card charges	0.50	0.50	0.40	0.20	0.10	0.40
12.5.3 Other services and professional fees	0.80	2.00	1.80	0.80	0.20	1.40
1-12 All expenditure groups	**349.80**	**414.90**	**392.00**	**243.50**	**159.40**	**348.30**

Note: The commodity and service categories are not comparable with those in publications before 2001-02
The numbering system is sequential, it does not use actual COICOP codes

2.3 Detailed household expenditure by age of household reference person (cont.) 2002-03

based on weighted data and including children's expenditure

Commodity or service	Under 30	30 and under 50	50 and under 65	65 and under 75	75 or over	All households
	Average weekly household expenditure (£)					
13 Other expenditure items	**50.20**	**82.10**	**59.30**	**27.40**	**17.80**	**57.90**
13.1 Housing: mortgage interest payments, water, council tax etc.	34.70	59.40	37.30	14.90	10.50	39.40
13.2 Licences, fines and transfers	2.60	3.00	3.50	2.00	1.20	2.70
13.3 Holiday spending	5.60	8.80	7.80	2.60	[0.90]	6.40
13.4 Money transfers and credit	7.30	10.80	10.70	7.80	5.20	9.40
13.4.1 Money, cash gifts given to children	[0.10]	0.40	[0.00]	..	..	0.20
13.4.2 Cash gifts and donations	4.70	8.10	9.20	7.30	5.10	7.50
13.4.3 Club instalment payments (child) and interest on credit cards	2.50	2.40	1.50	0.50	0.10	1.60
Total expenditure	**400.10**	**496.90**	**451.40**	**270.90**	**177.20**	**406.20**
14 Other items recorded						
14.1 Life assurance & contributions to pension funds	12.50	30.70	34.80	6.00	2.10	23.00
14.2 Other insurance inc. Friendly Societies	1.00	1.60	1.40	0.60	0.60	1.20
14.3 Income tax, payments less refunds	63.00	102.30	85.60	27.90	17.60	73.70
14.4 National insurance contributions	24.40	29.10	20.50	1.80	[0.90]	19.40
14.5 Purchase or alteration of dwellings, mortgages	18.80	48.90	32.80	13.80	7.10	31.90
14.6 Savings and investments	5.90	8.50	8.10	2.40	0.80	6.40
14.7 Pay off loan to clear other debt	5.10	3.80	1.90	[0.30]	..	2.50
14.8 Windfall receipts from gambling etc.	1.30	2.10	2.60	2.30	0.80	2.00

Note: The commodity and service categories are not comparable with those in publications before 2001-02
The numbering system is sequential, it does not use actual COICOP codes

2.4 Household expenditure by gross income quintile group where the household reference person is aged under 30

2001-02 – 2002-03

based on weighted data and including children's expenditure

	Lowest twenty per cent	Second quintile group	Third quintile group	Fourth quintile group	Highest twenty per cent	All house-holds
Lower boundary of group (£ per week)[1]		188	341	541	821	
Average number of grossed households (thousands)	490	450	580	590	460	2,570
Total number of households in sample (over 2 years)	309	279	341	321	206	1,456
Total number of persons in sample (over 2 years)	669	683	838	780	513	3,483
Total number of adults in sample (over 2 years)	360	435	596	656	466	2,513
Weighted average number of persons per household	2.1	2.4	2.4	2.3	2.6	2.4
Commodity or service			**Average weekly household expenditure (£)**			
1 Food & non-alcoholic drinks	24.40	26.60	30.60	32.90	41.10	31.10
2 Alcoholic drinks, tobacco & narcotics	8.10	10.50	10.80	11.50	12.40	10.70
3 Clothing & footwear	12.60	15.60	22.60	28.20	40.50	24.00
4 Housing[2], fuel & power	33.20	47.40	57.00	54.30	97.10	57.40
5 Household goods & services	14.00	19.40	20.50	30.60	37.70	24.40
6 Health	0.50	1.40	1.90	3.30	4.40	2.30
7 Transport	14.60	32.60	51.90	80.20	117.80	59.70
8 Communication	7.10	10.70	12.90	15.30	21.40	13.60
9 Recreation & culture	20.40	33.40	41.80	63.30	78.20	47.70
10 Education	6.40	2.40	3.80	2.00	5.60	4.00
11 Restaurants & hotels	15.40	27.10	38.40	51.20	78.50	42.30
12 Miscellaneous goods & services	11.30	19.40	31.50	38.30	51.90	30.70
1-12 All expenditure groups	167.90	246.60	323.80	411.10	586.50	348.00
13 Other expenditure items	7.10	27.50	48.90	67.20	95.90	49.80
Total expenditure	175.00	274.10	372.70	478.40	682.40	397.80
Average weekly expenditure per person (£)						
Total expenditure	82.30	116.20	155.90	204.10	264.90	168.70

Note: The commodity and service categories are not comparable with those in publications before 2001-02
1 Lower boundary of 2002-03 gross income quintile groups (£ per week)
2 Excluding mortgage interest payments, council tax and Northern Ireland rates

2.5 Household expenditure by gross income quintile group where the household reference person is aged 30 to 49 — 2001-02 – 2002-03

based on weighted data and including children's expenditure

	Lowest twenty per cent	Second quintile group	Third quintile group	Fourth quintile group	Highest twenty per cent	All house-holds
Lower boundary of group (£ per week)[1]		188	341	541	821	
Average number of grossed households (thousands)	1,010	1,210	2,070	2,460	2,750	9,510
Total number of households in sample (over 2 years)	633	774	1,294	1,483	1,577	5,761
Total number of persons in sample (over 2 years)	1,384	2,195	3,873	4,885	5,384	17,721
Total number of adults in sample (over 2 years)	780	1,183	2,243	2,982	3,478	10,666
Weighted average number of persons per household	2.1	2.7	2.9	3.2	3.3	3.0
Commodity or service		**Average weekly household expenditure (£)**				
1 Food & non-alcoholic drinks	27.60	35.80	41.70	50.90	62.10	47.80
2 Alcoholic drinks, tobacco & narcotics	9.20	11.00	12.40	14.30	15.30	13.20
3 Clothing & footwear	12.50	18.90	21.20	30.90	46.40	29.80
4 Housing[2], fuel & power	25.40	34.20	37.30	36.10	47.70	38.30
5 Household goods & services	15.60	18.40	26.70	33.50	60.40	36.00
6 Health	1.20	1.80	2.90	5.10	6.40	4.20
7 Transport	19.90	34.70	51.80	76.40	119.80	72.30
8 Communication	7.40	10.20	11.30	13.10	16.10	12.60
9 Recreation & culture	23.00	35.90	47.80	72.60	106.10	67.00
10 Education	1.60	2.00	3.40	5.20	19.20	8.10
11 Restaurants & hotels	14.40	22.70	30.30	44.30	69.40	42.60
12 Miscellaneous goods & services	13.60	19.50	28.70	40.90	64.50	39.50
1-12 All expenditure groups	171.50	244.90	315.30	423.30	633.40	411.30
13 Other expenditure items	17.40	37.60	57.80	84.00	144.20	82.80
Total expenditure	188.90	282.50	373.20	507.30	777.60	494.00
Average weekly expenditure per person (£)						
Total expenditure	89.70	103.30	129.40	158.40	233.50	165.00

Note: The commodity and service categories are not comparable with those in publications before 2001-02
1 Lower boundary of 2002-03 gross income quintile groups (£ per week)
2 Excluding mortgage interest payments, council tax and Northern Ireland rates

2.6 Household expenditure by gross income quintile group where the household reference person is aged 50 to 64

2001-02 – 2002-03

based on weighted data and including children's expenditure

	Lowest twenty per cent	Second quintile group	Third quintile group	Fourth quintile group	Highest twenty per cent	All house-holds
Lower boundary of group (£ per week)[1]		188	341	541	821	
Average number of grossed households (thousands)	920	1,090	1,270	1,360	1,450	6,090
Total number of households in sample (over 2 years)	587	676	754	792	767	3,576
Total number of persons in sample (over 2 years)	801	1,211	1,609	1,965	2,200	7,786
Total number of adults in sample (over 2 years)	769	1,106	1,475	1,740	1,973	7,063
Weighted average number of persons per household	1.4	1.8	2.2	2.5	2.9	2.2
Commodity or service	**Average weekly household expenditure (£)**					
1 Food & non-alcoholic drinks	**26.60**	**36.60**	**43.60**	**53.70**	**64.90**	**47.10**
2 Alcoholic drinks, tobacco & narcotics	**8.70**	**11.00**	**12.40**	**14.90**	**18.40**	**13.60**
3 Clothing & footwear	**6.70**	**13.20**	**19.70**	**26.60**	**41.20**	**23.20**
4 Housing[2], fuel & power	**22.40**	**30.30**	**33.60**	**34.50**	**43.50**	**33.90**
5 Household goods & services	**13.70**	**26.10**	**27.60**	**39.40**	**55.20**	**34.50**
6 Health	**1.90**	**4.40**	**5.50**	**6.90**	**13.10**	**6.90**
7 Transport	**23.60**	**33.60**	**58.70**	**76.80**	**127.00**	**69.20**
8 Communication	**5.50**	**7.60**	**9.10**	**11.40**	**15.90**	**10.40**
9 Recreation & culture	**22.30**	**36.80**	**55.70**	**74.50**	**106.00**	**63.50**
10 Education	**[0.20]**	**0.60**	**3.00**	**6.10**	**19.10**	**6.70**
11 Restaurants & hotels	**13.40**	**19.00**	**27.90**	**42.40**	**68.60**	**37.00**
12 Miscellaneous goods & services	**11.40**	**19.10**	**26.20**	**36.00**	**60.30**	**33.00**
1-12 All expenditure groups	**156.60**	**238.20**	**323.10**	**423.20**	**633.20**	**379.00**
13 Other expenditure items	**15.70**	**31.00**	**48.90**	**63.80**	**112.80**	**59.20**
Total expenditure	**172.20**	**269.20**	**372.00**	**487.00**	**746.00**	**438.20**
Average weekly expenditure per person (£)						
Total expenditure	**126.90**	**150.90**	**171.80**	**197.40**	**255.50**	**197.20**

Note: The commodity and service categories are not comparable with those in publications before 2001-02
1 Lower boundary of 2002-03 gross income quintile groups (£ per week)
2 Excluding mortgage interest payments, council tax and Northern Ireland rates

2.7 Household expenditure by gross income quintile group where the household reference person is aged 65 to 74

2001-02 – 2002-03

based on weighted data and including children's expenditure

	Lowest twenty per cent	Second quintile group	Third quintile group	Fourth quintile group	Highest twenty per cent	All house-holds
Lower boundary of group (£ per week)[1]		188	341	541	821	
Average number of grossed households (thousands)	1,010	1,180	590	320	160	3,260
Total number of households in sample (over 2 years)	630	722	363	179	92	1,986
Total number of persons in sample (over 2 years)	773	1,260	724	402	221	3,380
Total number of adults in sample (over 2 years)	764	1,245	712	389	213	3,323
Weighted average number of persons per household	1.2	1.7	2.0	2.3	2.4	1.7
Commodity or service			**Average weekly household expenditure (£)**			
1 Food & non-alcoholic drinks	**25.40**	**37.40**	**44.10**	**54.70**	**63.60**	**37.90**
2 Alcoholic drinks, tobacco & narcotics	**5.10**	**7.40**	**9.30**	**13.80**	**12.20**	**7.90**
3 Clothing & footwear	**5.90**	**10.60**	**17.00**	**23.40**	**25.30**	**12.30**
4 Housing[2], fuel & power	**23.00**	**28.50**	**28.90**	**38.60**	**51.80**	**29.00**
5 Household goods & services	**10.80**	**21.40**	**28.00**	**32.00**	**35.10**	**21.00**
6 Health	**1.80**	**4.80**	**9.70**	**8.00**	**9.60**	**5.30**
7 Transport	**11.00**	**30.50**	**47.70**	**65.80**	**95.30**	**34.20**
8 Communication	**5.00**	**6.30**	**7.10**	**9.90**	**12.40**	**6.70**
9 Recreation & culture	**18.30**	**38.00**	**52.50**	**75.10**	**93.60**	**40.90**
10 Education	..	..	..	..	..	**0.70**
11 Restaurants & hotels	**7.40**	**14.90**	**26.70**	**40.20**	**51.20**	**19.00**
12 Miscellaneous goods & services	**9.40**	**18.30**	**29.60**	**36.40**	**58.10**	**21.30**
1-12 All expenditure groups	**123.00**	**218.10**	**301.10**	**398.80**	**518.00**	**236.40**
13 Other expenditure items	**13.00**	**24.40**	**34.40**	**45.90**	**88.70**	**27.70**
Total expenditure	**136.00**	**242.50**	**335.40**	**444.70**	**606.70**	**264.10**
Average weekly expenditure per person (£)						
Total expenditure	**110.70**	**138.60**	**168.40**	**195.40**	**250.90**	**153.80**

Note: The commodity and service categories are not comparable with those in publications before 2001-02
1 Lower boundary of 2002-03 gross income quintile groups (£ per week)
2 Excluding mortgage interest payments, council tax and Northern Ireland rates

2.8 Household expenditure by gross income quintile group where the household reference person is aged 75 or over

2001-02 – 2002-03

based on weighted data and including children's expenditure

	Lowest twenty per cent	Second quintile group	Third quintile group	Fourth quintile group	Highest twenty per cent	All house-holds
Lower boundary of group (£ per week)[1]		188	341	541	821	
Average number of grossed households (thousands)	1,450	960	360	140	50	2,970
Total number of households in sample (over 2 years)	793	531	196	75	26	1,621
Total number of persons in sample (over 2 years)	923	865	357	140	53	2,338
Total number of adults in sample (over 2 years)	922	862	354	140	53	2,331
Weighted average number of persons per household	1.2	1.6	1.8	1.9	2.3	1.4
Commodity or service	**Average weekly household expenditure (£)**					
1 **Food & non-alcoholic drinks**	21.30	32.00	39.00	42.20	57.70	28.60
2 **Alcoholic drinks, tobacco & narcotics**	2.70	5.60	7.20	10.20	13.70	4.70
3 **Clothing & footwear**	4.50	7.70	14.30	11.20	[33.10]	7.60
4 **Housing[2], fuel & power**	22.80	26.20	31.50	35.80	35.00	25.80
5 **Household goods & services**	9.80	19.30	20.90	37.10	32.50	16.00
6 **Health**	2.10	3.20	5.70	5.60	[7.20]	3.10
7 **Transport**	5.50	17.30	32.50	43.70	65.80	15.60
8 **Communication**	4.10	5.00	6.30	7.60	11.90	5.00
9 **Recreation & culture**	12.90	23.20	31.40	43.00	78.80	21.20
10 **Education**	..	..	..	..	..	[0.40]
11 **Restaurants & hotels**	5.70	10.90	17.20	28.20	[39.10]	10.50
12 **Miscellaneous goods & services**	9.80	16.00	27.00	42.20	127.10	17.40
1-12 **All expenditure groups**	101.40	166.30	233.00	309.70	514.40	155.90
13 **Other expenditure items**	10.40	19.10	29.20	34.30	87.30	17.90
Total expenditure	111.80	185.40	262.20	344.00	601.80	173.80
Average weekly expenditure per person (£)						
Total expenditure	96.90	115.20	144.20	185.70	267.30	121.10

Note: The commodity and service categories are not comparable with those in publications before 2001-02

1 Lower boundary of 2002-03 gross income quintile groups (£ per week)

2 Excluding mortgage interest payments, council tax and Northern Ireland rates

Chapter 3
Expenditure by socio-economic characteristics

- In 2002-03 households with a reference person in **full-time employment** had the highest average expenditure at £540 a week. This was almost two and a half times the expenditure of households where the reference person was **retired** (£223 a week).

- For households where the reference person was **in employment** spending was greatest on transport and recreation and culture at £78 and £71 a week respectively. Where the household reference person was **unemployed**, transport, food and housing were the largest items of expenditure (£44, £33 and £33 a week respectively). For **economically inactive** households, the highest expenditure categories were food and recreation and culture, both around £35 a week.

- Total expenditure increased by age at which the household reference person completed continuous full-time education from **£210** a week for those who were aged 14 or under, to nearly **£605** a week for those aged 22 or over. Those who left school aged 14 and under had an average age of 73, so most would be retired, and those who left full-time education aged 15 were on average aged 57 and nearing retirement age.

- Expenditure on most commodities and services increased with age at which the reference person completed continuous full-time education. Exceptions were expenditure on **alcoholic drink and tobacco**, which was similar across all groups except those where the HRP completed education aged under 14.

3.1 Expenditure by economic activity status of the HRP

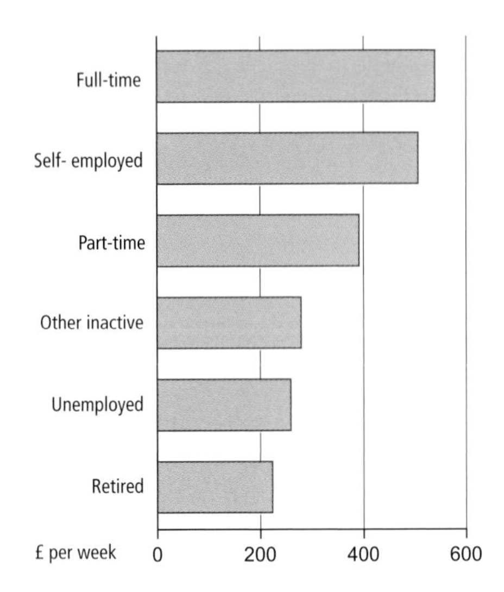

Chapter 3 Expenditure by socio-economic characteristics

This chapter contains tables showing household expenditure analysed by some socio-economic characteristics of the household reference person (HRP). **Tables 3.2** and **3.3** also include a further breakdown by income for those in full-time employment and those who are self-employed. Definitions for all categories can be found in Appendix D.

Economic activity status

Table 3.1 and **Figure 3.1** show household expenditure by the economic activity status of the HRP. Households with an HRP in full-time employment had the highest total expenditure, £540 a week, compared with £391 for part-time employees and £507 for the self-employed. Households with an unemployed reference person spent £259 a week, while households with a retired reference person spent the least, £223 a week – almost two and a half times less than those in full-time employment.

Transport and recreation and culture were the greatest expenditure items for households with an HRP in full-time employment, at £83 and £74 respectively. For households where the reference person was unemployed, the largest expenditure items were transport at £44 a week, and food and non-alcoholic drinks at £33 a week. Households where the HRP was economically inactive spent the highest proportion on food and non-alcoholic drinks at 15 per cent or £35 a week, compared to nine per cent for households with an HRP in full-time employment.

Table 3.2 shows household expenditure by income quintile group for full-time employees. Transport was the highest expenditure category for the third income quintile and above, while for the two lowest groups, it was housing, fuel and power. Households in the highest income group spent over six times as much on transport than those in the lowest group.

Table 3.3 shows expenditure by income quintile group for households with a self-employed reference person. Expenditure increased generally with income group, with the highest category of expenditure for all groups being transport followed by recreation and culture.

Table 3.4 and **figure 3.2** show expenditure on the main commodities generally increases with the number of persons working. Transport was the highest expenditure item for all households except for those with no persons working. For this group 15 per cent of their total expenditure was on food and non-alcoholic drinks, higher than for any other group. This group also spent a higher proportion on housing, fuel and power (12 per cent) than other households. In absolute terms, households with four or more working persons spent around six times as much on clothing and footwear, transport and restaurants and hotels than those with no persons working.

Table 3.5 and **Figure 3.3** show households by age at which the HRP completed continuous full-time education. Expenditure increased with time spent in full-time education, from £210 a week, for households whose reference person left aged 14 or under, to £600 a week for those who left aged 22 or over. Those who left school aged 14 and under had an average age of 73, so many would have been retired. Their highest categories of expenditure were food and non-alcoholic drinks, housing, fuel and power and recreation and culture. By contrast, for households whose HRP left full-time education aged 22 or over, the highest spending category was transport. The amount spent on alcohol and tobacco was similar for all groups at around £12 except for those where the HRP finished school aged 14 or under, who spent half that amount.

3.2 Proportion spent on transport, food and housing by number of persons working

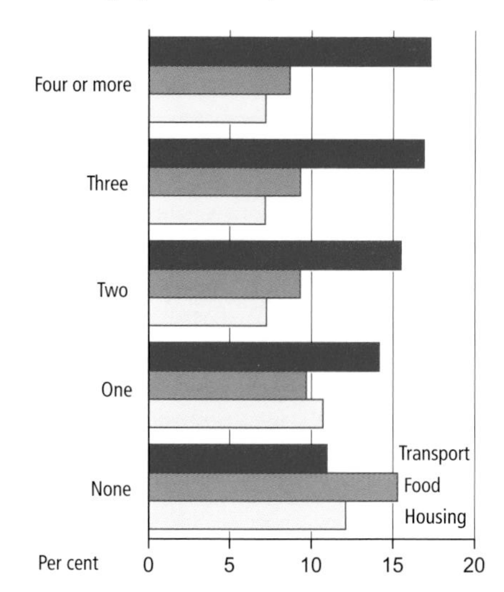

3.3 Expenditure on selected items by age HRP completed continuous full-time education

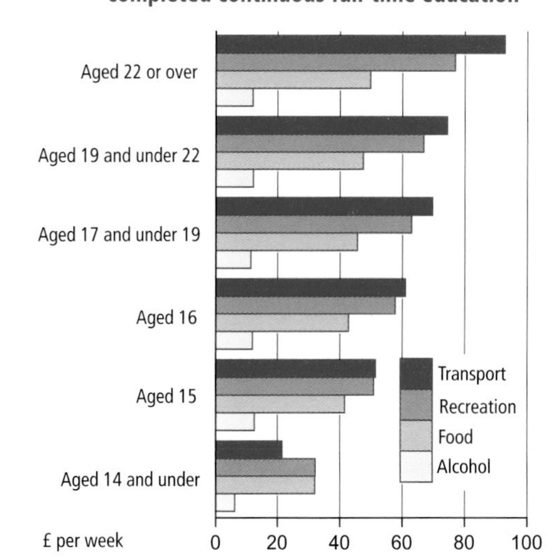

3.1 Household expenditure by economic activity status of household reference person

2002-03

based on weighted data and including children's expenditure

	Employees			Self-employed	All in employment[1]
	Full-time	Part-time	All		
Grossed number of households (thousands)	10,750	1,920	12,670	1,920	14,630
Total number of households in sample	2,985	552	3,537	542	4,090
Total number of persons in sample	8,173	1,427	9,600	1,501	11,124
Total number of adults in sample	5,902	942	6,844	1,070	7,928
Weighted average number of persons per household	2.7	2.6	2.7	2.7	2.7
Commodity or service	**Average weekly household expenditure (£)**				
1 Food & non-alcoholic drinks	48.30	42.00	47.40	51.40	47.80
2 Alcoholic drinks, tobacco & narcotics	13.60	11.20	13.20	13.40	13.20
3 Clothing & footwear	29.80	25.00	29.10	27.70	28.90
4 Housing[2], fuel & power	42.40	45.50	42.90	42.60	42.80
5 Household goods & services	38.90	25.10	36.80	38.30	36.90
6 Health	5.60	3.20	5.20	5.80	5.30
7 Transport	82.90	53.80	78.50	78.70	78.30
8 Communication	12.70	12.10	12.60	14.90	12.90
9 Recreation & culture	74.00	58.70	71.70	66.50	70.80
10 Education	8.20	4.30	7.60	9.50	7.80
11 Restaurants & hotels	49.70	36.40	47.70	43.10	47.10
12 Miscellaneous goods & services	44.70	29.90	42.50	38.30	41.90
1-12 All expenditure groups	450.80	347.20	435.10	430.30	433.70
13 Other expenditure items	88.90	44.10	82.10	76.70	81.20
Total expenditure	539.70	391.30	517.20	507.00	514.90
Average weekly expenditure per person (£)					
Total expenditure	200.10	149.70	192.70	184.80	191.30

Note: The commodity and service categories are not comparable with those in publications before 2001-02

1 Includes households where the head was on a government-supported training scheme.

2 Excluding mortgage interest payments, council tax and Northern Ireland rates

3.1 Household expenditure by economic activity status of household reference person (cont.) 2002-03

based on weighted data and including children's expenditure

	ILO unem-ployed	All economi-cally active[1]	Economically inactive			All house-holds
			Retired	Other	All	
Grossed number of households (thousands)	490	15,120	6,220	3,000	9,220	24,340
Total number of households in sample	139	4,229	1,804	892	2,696	6,925
Total number of persons in sample	357	11,481	2,856	2,242	5,098	16,579
Total number of adults in sample	228	8,156	2,825	1,465	4,290	12,446
Weighted average number of persons per household	2.6	2.7	1.6	2.5	1.9	2.4
Commodity or service	**Average weekly household expenditure (£)**					
1 Food & non-alcoholic drinks	**33.40**	**47.40**	**34.10**	**37.30**	**35.10**	**42.70**
2 Alcoholic drinks, tobacco & narcotics	**8.30**	**13.10**	**7.00**	**11.80**	**8.50**	**11.40**
3 Clothing & footwear	**17.70**	**28.50**	**9.70**	**17.10**	**12.10**	**22.30**
4 Housing[2], fuel & power	**33.30**	**42.50**	**26.90**	**29.30**	**27.70**	**36.90**
5 Household goods & services	**23.40**	**36.50**	**19.00**	**22.10**	**20.00**	**30.20**
6 Health	**1.30**	**5.20**	**4.60**	**3.70**	**4.30**	**4.80**
7 Transport	**43.60**	**77.20**	**25.10**	**39.00**	**29.60**	**59.20**
8 Communication	**8.60**	**12.80**	**6.00**	**9.20**	**7.10**	**10.60**
9 Recreation & culture	**27.70**	**69.50**	**32.90**	**39.20**	**35.00**	**56.40**
10 Education	**[1.80]**	**7.60**	**0.70**	**2.20**	**1.20**	**5.20**
11 Restaurants & hotels	**18.30**	**46.10**	**15.30**	**23.30**	**17.90**	**35.40**
12 Miscellaneous goods & services	**16.00**	**41.00**	**19.50**	**21.30**	**20.10**	**33.10**
1-12 All expenditure groups	**233.50**	**427.30**	**200.90**	**255.50**	**218.60**	**348.30**
13 Other expenditure items	**25.50**	**79.40**	**22.20**	**23.70**	**22.70**	**57.90**
Total expenditure	**259.00**	**506.70**	**223.10**	**279.20**	**241.30**	**406.20**
Average weekly expenditure per person (£)						
Total expenditure	**99.90**	**188.50**	**140.90**	**112.00**	**128.50**	**170.60**

Note: The commodity and service categories are not comparable with those in publications before 2001-02
1 Includes households where the head was on a government-supported training scheme.
2 Excluding mortgage interest payments, council tax and Northern Ireland rates

3.2 Household expenditure by gross income: the household reference person is a full-time employee

2002-03

based on weighted data and including children's expenditure

		Lowest twenty per cent	Second quintile group	Third quintile group	Fourth quintile group	Highest twenty per cent	All house-holds
Lower boundary of group (£ per week)			188	341	541	821	
Grossed number of households (thousands)		90	1,000	2,540	3,420	3,700	10,750
Total number of households in sample		27	286	727	964	981	2,985
Total number of persons in sample		37	489	1,777	2,752	3,118	8,173
Total number of adults in sample		32	393	1,262	1,935	2,280	5,902
Weighted average number of persons per household		1.3	1.7	2.4	2.8	3.1	2.7
Commodity or service			**Average weekly household expenditure (£)**				
1	**Food & non-alcoholic drinks**	**19.50**	**27.90**	**37.70**	**47.90**	**62.10**	**48.30**
2	**Alcoholic drinks, tobacco & narcotics**	**15.70**	**11.40**	**11.00**	**13.40**	**16.00**	**13.60**
3	**Clothing & footwear**	**[8.30]**	**12.10**	**19.00**	**29.10**	**43.20**	**29.80**
4	**Housing[1], fuel & power**	**41.00**	**39.90**	**38.00**	**36.90**	**51.30**	**42.40**
5	**Household goods & services**	**[7.70]**	**16.80**	**24.10**	**32.90**	**61.20**	**38.90**
6	**Health**	**[0.60]**	**2.10**	**3.60**	**5.50**	**8.00**	**5.60**
7	**Transport**	**20.00**	**32.40**	**51.60**	**78.50**	**123.40**	**82.90**
8	**Communication**	**6.90**	**8.80**	**10.00**	**11.90**	**16.50**	**12.70**
9	**Recreation & culture**	**28.80**	**32.20**	**47.20**	**74.30**	**104.40**	**74.00**
10	**Education**	**..**	**[1.10]**	**3.80**	**3.50**	**17.40**	**8.20**
11	**Restaurants & hotels**	**[11.30]**	**23.50**	**31.20**	**44.10**	**75.60**	**49.70**
12	**Miscellaneous goods & services**	**30.20**	**19.80**	**30.50**	**41.00**	**64.90**	**44.70**
1-12	**All expenditure groups**	**197.40**	**228.00**	**307.80**	**419.20**	**644.00**	**450.80**
13	**Other expenditure items**	**33.90**	**42.10**	**57.70**	**78.00**	**134.20**	**88.90**
Total expenditure		**231.30**	**270.10**	**365.60**	**497.10**	**778.20**	**539.70**
Average weekly expenditure per person (£)							
Total expenditure		**172.20**	**160.10**	**152.80**	**179.10**	**248.10**	**200.10**

Note: The commodity and service categories are not comparable with those in publications before 2001-02
1 Excluding mortgage interest payments, council tax and Northern Ireland rates

3.3 Household expenditure by gross income: the household reference person is self-employed

2001-02 – 2002-03

based on weighted data and including children's expenditure

	Lowest twenty per cent	Second quintile group	Third quintile group	Fourth quintile group	Highest twenty per cent	All house-holds
Lower boundary of group (£ per week)[1]		188	341	541	821	
Average number of grossed households (thousands)	180	310	430	400	520	1,840
Total number of households in sample (over 2 years)	110	187	259	246	287	1,089
Total number of persons in sample (over 2 years)	202	458	744	767	884	3,055
Total number of adults in sample (over 2 years)	155	337	508	532	644	2,176
Weighted average number of persons per household	1.7	2.4	2.8	3.0	3.0	2.7
Commodity or service			**Average weekly household expenditure (£)**			
1 Food & non-alcoholic drinks	33.10	38.90	49.30	60.90	60.60	51.70
2 Alcoholic drinks, tobacco & narcotics	10.80	10.10	11.70	15.30	16.40	13.50
3 Clothing & footwear	14.30	18.70	22.90	30.50	41.50	28.30
4 Housing[2], fuel & power	27.50	35.80	41.80	38.10	54.10	42.20
5 Household goods & services	27.80	26.20	31.30	45.30	61.10	41.80
6 Health	1.90	4.20	3.60	8.20	6.10	5.20
7 Transport	44.40	55.30	63.60	89.90	125.50	83.60
8 Communication	9.20	10.90	14.30	15.80	18.20	14.70
9 Recreation & culture	34.60	47.00	57.20	68.60	101.60	68.40
10 Education	..	1.00	3.90	12.10	30.30	12.70
11 Restaurants & hotels	18.50	29.90	34.60	47.90	73.10	46.00
12 Miscellaneous goods & services	22.80	25.80	30.90	44.20	61.10	40.70
1-12 All expenditure groups	248.80	303.70	365.00	476.80	649.60	448.60
13 Other expenditure items	41.90	51.80	66.30	86.30	166.60	93.70
Total expenditure	290.70	355.50	431.30	563.10	816.10	542.30
Average weekly expenditure per person (£)						
Total expenditure	168.40	150.30	154.40	187.90	267.60	198.40

Note: The commodity and service categories are not comparable with those in publications before 2001-02
1 Lower boundaryof 2002-03 gross income quintile groups (£ per week)
2 Excluding mortgage interest payments, council tax and Northern Ireland rates

3.4 Household expenditure by number of persons working — 2002-03

based on weighted data and including children's expenditure

	Number of persons working					All
	None	One	Two	Three	Four or more	house-holds
Grossed number of households (thousands)	8,430	6,650	7,350	1,450	470	24,350
Total number of households in sample	2,498	1,916	2,066	341	106	6,927
Total number of persons in sample	4,403	4,219	6,193	1,284	487	16,586
Total number of adults in sample	3,679	3,014	4,350	1,003	404	12,450
Weighted average number of persons per household	1.7	2.2	2.9	3.7	4.5	2.4
Weighted average age of head of household	65	45	42	47	48	51
Employment status of head[1]:						
- % working full-time or self-employed	*0*	*67*	*89*	*89*	*89*	*51*
- % working part-time	*0*	*18*	*8*	*7*	*10*	*8*
- % not working	*100*	*14*	*2*	*3*	*1*	*41*
Commodity or service	**Average weekly household expenditure (£)**					
1 Food & non-alcoholic drinks	**32.60**	**38.10**	**52.50**	**63.40**	**74.40**	**42.70**
2 Alcoholic drinks, tobacco & narcotics	**7.60**	**11.00**	**14.30**	**16.80**	**21.50**	**11.40**
3 Clothing & footwear	**10.50**	**18.70**	**32.10**	**44.60**	**62.20**	**22.30**
4 Housing[2], fuel & power	**25.80**	**42.00**	**40.90**	**48.80**	**61.80**	**36.90**
5 Household goods & services	**18.40**	**29.80**	**40.10**	**44.00**	**52.10**	**30.20**
6 Health	**4.10**	**4.10**	**5.70**	**7.50**	**6.00**	**4.80**
7 Transport	**23.40**	**55.50**	**87.10**	**114.50**	**147.90**	**59.20**
8 Communication	**6.30**	**10.70**	**13.00**	**18.60**	**24.40**	**10.60**
9 Recreation & culture	**30.20**	**50.90**	**79.40**	**94.00**	**126.90**	**56.40**
10 Education	**1.20**	**5.40**	**8.50**	**10.70**	**[6.90]**	**5.20**
11 Restaurants & hotels	**14.90**	**33.00**	**49.20**	**76.20**	**97.80**	**35.40**
12 Miscellaneous goods & services	**18.20**	**29.60**	**47.50**	**53.50**	**62.10**	**33.10**
1-12 All expenditure groups	**193.20**	**328.80**	**470.30**	**592.60**	**743.90**	**348.30**
13 Other expenditure items	**19.40**	**61.70**	**90.30**	**83.30**	**109.90**	**57.90**
Total expenditure	**212.50**	**390.50**	**560.50**	**676.00**	**853.80**	**406.20**
Average weekly expenditure per person (£)						
Total expenditure	**123.10**	**180.80**	**190.60**	**183.80**	**188.30**	**170.50**

Note: The commodity and service categories are not comparable with those in publications before 2001-02

1 Excludes households where the head was on a government-supported training scheme.

2 Excluding mortgage interest payments, council tax and Northern Ireland rates

3.5 Household expenditure by age at which the household reference person completed continuous full-time education 2002-03

based on weighted data and including children's expenditure

		Aged 14 and under	Aged 15	Aged 16	Aged 17 and under 19	Aged 19 and under 22	Aged 22 or over
Grossed number of households (thousands)		3,380	5,160	7,260	3,920	2,220	2,260
Total number of households in sample		998	1,442	2,092	1,137	619	605
Total number of persons in sample		1,706	3,238	5,587	2,893	1,552	1,517
Total number of adults in sample		1,612	2,716	3,736	2,045	1,120	1,132
Weighted average number of persons per household		1.7	2.3	2.6	2.5	2.5	2.5
Weighted average age of head of household		73	57	45	46	44	43
Commodity or service		**Average weekly household expenditure (£)**					
1	**Food & non-alcoholic drinks**	**32.30**	**41.80**	**43.00**	**45.80**	**47.70**	**50.00**
2	**Alcoholic drinks, tobacco & narcotics**	**6.40**	**12.70**	**12.10**	**11.50**	**12.30**	**12.20**
3	**Clothing & footwear**	**10.10**	**17.90**	**24.50**	**26.70**	**31.30**	**26.40**
4	**Housing[1], fuel & power**	**27.50**	**29.90**	**36.10**	**39.60**	**45.60**	**50.40**
5	**Household goods & services**	**17.50**	**26.90**	**29.60**	**32.40**	**37.50**	**48.90**
6	**Health**	**4.00**	**4.20**	**4.20**	**5.50**	**5.80**	**7.40**
7	**Transport**	**21.80**	**51.70**	**61.20**	**69.90**	**74.60**	**93.00**
8	**Communication**	**5.80**	**9.00**	**11.60**	**11.40**	**12.70**	**14.00**
9	**Recreation & culture**	**32.40**	**51.10**	**57.90**	**63.20**	**67.00**	**77.10**
10	**Education**	**[0.90]**	**1.90**	**3.50**	**6.30**	**9.50**	**16.60**
11	**Restaurants & hotels**	**15.00**	**29.20**	**36.70**	**42.50**	**43.90**	**53.40**
12	**Miscellaneous goods & services**	**16.20**	**25.50**	**32.80**	**39.60**	**42.60**	**56.10**
1-12	**All expenditure groups**	**189.80**	**301.90**	**353.40**	**394.70**	**430.50**	**505.60**
13	**Other expenditure items**	**20.00**	**41.10**	**58.70**	**69.80**	**91.80**	**99.30**
Total expenditure		**209.70**	**343.00**	**412.10**	**464.50**	**522.30**	**604.90**
Average weekly expenditure per person (£) **Total expenditure**		**123.50**	**150.10**	**156.60**	**184.70**	**211.50**	**244.20**

Note: The commodity and service categories are not comparable with those in publications before 2001-02
1 Excluding mortgage interest payments, council tax and Northern Ireland rates

3.6 Household expenditure by socio-economic class[2] of the household reference person

2002-03

based on weighted data and including children's expenditure

		Higher managerial and professional			Lower manag-erial & profess-ional	Interm-ediate
		Large employers & higher managerial	Higher profess-ional	All		
Grossed number of households (thousands)		1,070	1,590	2,670	4,220	1,440
Total number of households in sample		287	420	707	1,174	419
Total number of persons in sample		846	1,117	1,963	3,161	1,034
Total number of adults in sample		579	827	1,406	2,270	738
Weighted average number of persons per household		2.9	2.6	2.7	2.6	2.5
Commodity or service		**Average weekly household expenditure (£)**				
1	Food & non-alcoholic drinks	58.50	52.30	54.80	50.50	39.40
2	Alcoholic drinks, tobacco & narcotics	13.20	14.60	14.10	13.00	11.00
3	Clothing & footwear	41.40	31.00	35.20	32.00	24.60
4	Housing, fuel & power[3]	55.90	52.10	53.60	40.30	38.30
5	Household goods & services	69.00	49.60	57.40	41.50	27.10
6	Health	8.20	7.80	8.00	6.60	5.40
7	Transport	123.10	102.10	110.50	86.20	68.70
8	Communication	14.40	13.90	14.10	13.50	10.60
9	Recreation & culture	106.30	89.60	96.30	76.30	58.60
10	Education	21.00	14.60	17.20	9.30	4.40
11	Restaurants & hotels	67.80	63.60	65.30	52.40	38.20
12	Miscellaneous goods & services	61.30	53.80	56.80	49.80	36.50
1-12	All expenditure groups	640.10	545.00	583.20	471.40	362.80
13	Other expenditure items	137.00	111.40	121.70	96.80	72.20
Total expenditure		777.10	656.40	704.90	568.10	434.90
Average weekly expenditure per person (£)						
Total expenditure		265.40	251.60	257.50	215.10	177.40

Note: The commodity and service categories are not comparable with those in publications before 2001-02
1 Includes those who have never worked and full-time students
2 Excludes those who are economically inactive
3 Excluding mortgage interest payments, council tax and Northern Ireland rates

3.6 Household expenditure by socio-economic class[2] of the household reference person (cont.)

2002-03

based on weighted data and including children's expenditure

		Small employers	Lower supervisory	Semi-routine	Routine	Long-term unemployed[1]	All households
Grossed number of households (thousands)		1,380	1,640	1,810	1,640	340	24,350
Total number of households in sample		396	464	512	472	91	6,927
Total number of persons in sample		1,131	1,310	1,374	1,297	223	16,586
Total number of adults in sample		787	944	936	910	172	12,450
Weighted average number of persons per household		2.8	2.8	2.7	2.7	2.5	2.4
Commodity or service		**Average weekly household expenditure (£)**					
1	Food & non-alcoholic drinks	50.40	46.90	41.10	42.10	30.90	42.70
2	Alcoholic drinks, tobacco & narcotics	12.30	14.30	12.50	14.50	9.00	11.40
3	Clothing & footwear	25.90	25.10	23.90	23.40	24.50	22.30
4	Housing, fuel & power[3]	40.60	33.50	41.10	40.40	67.20	36.90
5	Household goods & services	32.40	31.60	23.10	25.70	13.90	30.20
6	Health	3.70	3.90	2.30	3.00	2.00	4.80
7	Transport	75.40	65.00	55.10	52.80	37.80	59.20
8	Communication	14.20	13.20	10.40	10.60	17.10	10.60
9	Recreation & culture	59.80	68.10	52.20	51.20	44.00	56.40
10	Education	5.80	3.70	2.10	0.70	14.80	5.20
11	Restaurants & hotels	38.00	41.50	32.60	32.40	42.90	35.40
12	Miscellaneous goods & services	34.30	36.40	27.70	26.20	18.20	33.10
1-12	All expenditure groups	392.80	383.10	324.10	322.90	322.30	348.30
13	Other expenditure items	66.50	65.90	46.70	44.20	17.20	57.90
Total expenditure		459.30	449.00	370.80	367.10	339.50	406.20
Average weekly expenditure per person (£) Total expenditure		161.60	159.50	138.60	135.30	134.50	170.50

Note: The commodity and service categories are not comparable with those in publications before 2001-02
1 Includes those who have never worked and full-time students
2 Excludes those who are economically inactive
3 Excluding mortgage interest payments, council tax and Northern Ireland rates

Chapter 4

Expenditure by household composition, income & tenure

Income is not adjusted to take into account the different composition of households (equivalisation) as done in some other income analyses

- Average weekly expenditure was highest for households consisting of three or more adults with children, at £634 a week. However, one person non-retired households spent the most per person at £268 a week. Households with two adults and two children spent an average of £585 a week.

- Expenditure was lowest for one person households mainly dependent on a state pension, at £104 a week.

- One person non-retired households had an average weekly spend of £268. The corresponding figures for one person retired households, dependent and non-dependent on a state pension, were £104 and £185 respectively.

Households with children

- Average weekly expenditure by households with children increased as the number of adults in the household increased, from around £260 a week for one adult households to £634 a week where there were three or more adults.

- The highest expenditure categories for single adult households with children were food, housing and recreation. The highest categories for two-adult households with children were recreation, transport and food.

4.1 Expenditure by one person retired and non-retired households

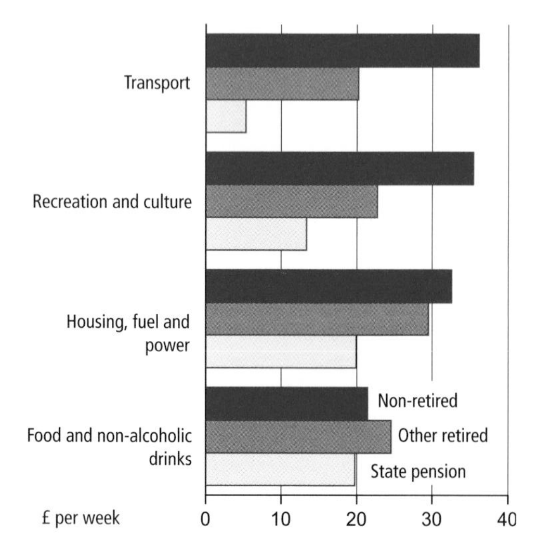

Chapter 4
Expenditure by household composition, income and tenure

Chapter 4 contains tables analysing expenditure by household composition and income. **Table 4.1** shows spending patterns for a selection of household compositions. **Tables 4.2 to 4.9** provide breakdowns by household composition and income group, and are based on two years data (2001-02 and 2002-03).

Household composition

Table 4.1 shows that average weekly expenditure was highest for households comprising three or more adults. These households spent over £600 a week, including over £100 on transport and over £80 on recreation and culture. In contrast, retired one person households mainly dependent on a state pension spent only £100 a week, of which around £40 was spent on food and housing and only £5 on transport.

Figure 4.1 compares expenditure on selected items by one person retired and one person non-retired households. In general, non-retired households spent more than retired households. Amongst retired households, those not dependent on a state pension generally spent more than those reliant on a state pension. The degree to which this was the case varied with the items. Weekly expenditure on food and non-alcoholic drinks was fairly similar between the groups, ranging from £19.60 to £24.40. One person retired households who were dependent on a state pension spent approximately £10 more a week on housing and recreation than those reliant on a state pension. Transport was the item of expenditure that varied the most between the three groups. One person retired households dependent on a state pension spent only £5.20 on transport whilst those not reliant on a pension spent £20.10. Single non-retired households spent £36.00 a week on transport.

Expenditure by household composition and income

Tables 4.2 to 4.9 show expenditure by income group for different household composition types. Many of the differences in spending between different types of household are the result of differences in income. These tables allow households in the same income groups but with different characteristics to be compared. The analysis in this section concentrates on households with the lowest and middle fifths of incomes. Please note that figures are not shown for certain income composition groups. This is because very few households fall into these categories.

Variations between different household types in the lowest and middle income quintiles

In the lowest income quintile, the highest item of expenditure for most households types was food and non-alcoholic drinks. One person retired households dependent on a state pension and retired couples spent around 20 per cent of total expenditure on this category. For one person retired households not dependent on a state pension and one person non-retired households, the highest spending category was housing, fuel and power. The only household type for whom transport was the highest expenditure category was one man, one woman non-retired households. Households in this group spent £40.50 a week.

Alcoholic drinks and tobacco

Figure 4.2a compares the proportion of total expenditure on alcoholic drinks and tobacco for different household types in the lowest income group . One person non-retired households had the highest proportion of spending on this item at 5.5 per cent followed by two adult and one adult households both with children at 4.9 per cent and 4.8 per cent respectively.

Figure 4.2b provides a comparison of expenditure on this item by the lowest and third quintiles. For all household types, expenditure did not vary much between the lowest and third quintiles. Households consisting of two adults with children spent the most within both the lowest and third quintiles at £13.90 and £13.70 per week respectively. Although one person non-retired households in the lowest quintile spent a relatively high proportion on this item, the absolute expenditure amount was only £7.70 per week. Conversely, non-retired couples in the lowest quintile spent £10.90 per week on alcoholic drinks and tobacco but this accounted for only 4.1 per cent of their total expenditure. One person retired households not dependent on a state pension spent the least, proportionately and in absolute terms, on alcoholic drinks and tobacco.

4.2a Proportion spent on alcohol and tobacco: lowest income group

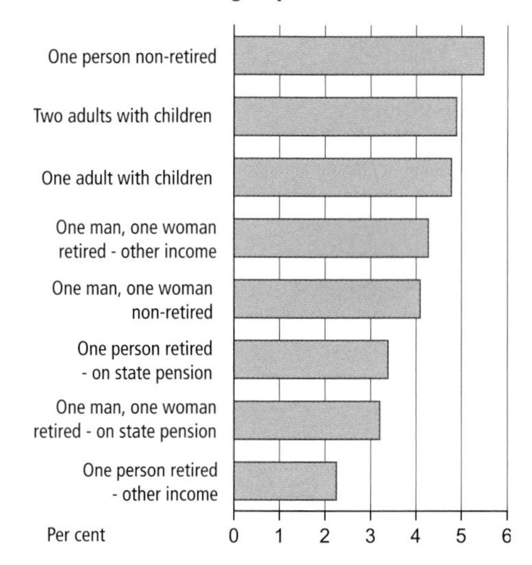

4.2b Expenditure on alcohol and tobacco: lowest and third income groups

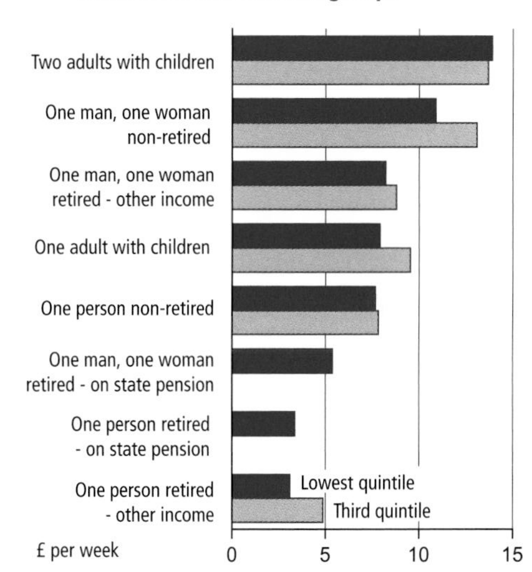

4.3a Proportion spent on transport: lowest income group

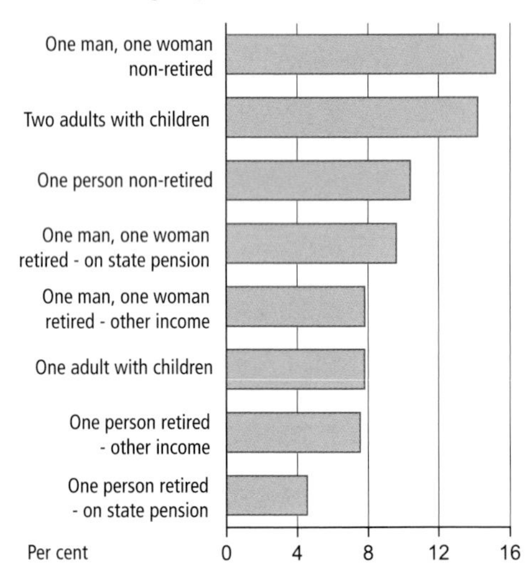

4.3b Expenditure on transport: lowest and third income groups

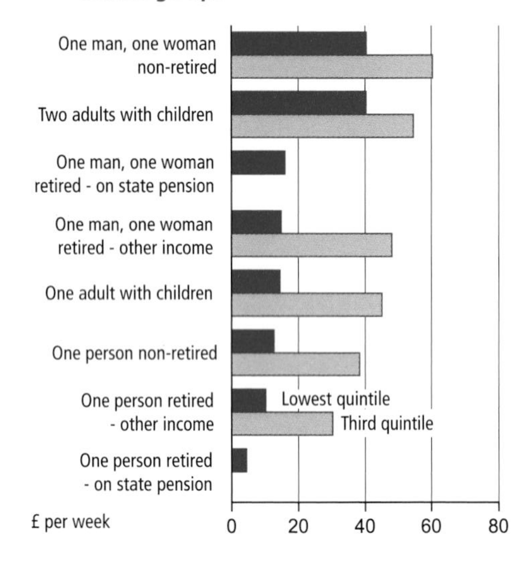

Transport

Figure 4.3a shows the proportion of total expenditure spent on transport for different household compositions in the lowest income group. Non-retired couples spent the highest proportion of their total expenditure on transport at 15.2 per cent followed by households with two adults with children at 14.2 per cent.

Figure 4.3b provides a comparison of expenditure on transport by the lowest and third income groups. Within the lowest income quintile, non-retired couples and households with two adults with children spent the most on transport at over £40 per week. Similarly, these two groups spent the most on transport within the third quintile. For the remaining household types, households in the third income group spent around three times more on transport than those in the lowest income group. One person retired households dependent on a state pension spent the least on transport, both proportionately and in absolute terms.

Tenure

Table 4.10 and **Figure 4.4** show average weekly household expenditure by tenure type. At £463 a week, total expenditure by owner occupiers was more than double that of social renters. Mortgaged households were by far the highest spenders with an average weekly spend of £545, the largest areas of spending being on transport and recreation and culture.

Figure 4.5 shows the proportion of total expenditure spent on selected items by tenure type. Social renters spent 11 per cent of their money on transport compared to 18 per cent on housing, fuel and power. Conversely, owners spent 15 per cent on transport and only 6 per cent on housing, fuel and power. The proportion spent on recreation and culture was similar between all three groups ranging from 12 per cent for private renters to 14 per cent for owners. Whilst social renters spent on average nearly £15 a week less on food and non-alcoholic drinks than home owners, this represented a significantly higher proportion of their total weekly spend.

4.4 Expenditure by tenure

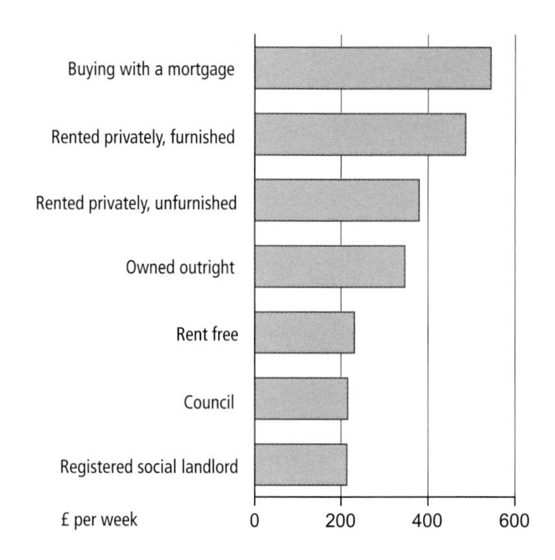

4.5 Proportion spent on selected items by tenure

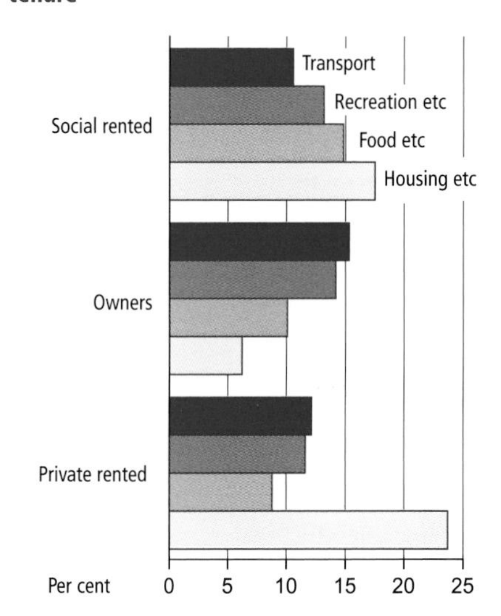

4.1 Expenditure by household composition 2002-03

based on weighted data and including children's expenditure

	Retired households				Non-retired	
	State pension[1]		Other retired			
	One person	One man and one woman	One person	One man and one woman	One person	One man and one woman
Grossed number of households (thousands)	1,640	700	1,470	1,900	3,490	5,020
Total number of households in sample	459	219	416	580	978	1,399
Total number of persons in sample	459	438	416	1,160	978	2,798
Total number of adults in sample	459	438	416	1,160	978	2,798
Weighted average number of persons per household	1.0	2.0	1.0	2.0	1.0	2.0
Commodity or service	**Average weekly household expenditure (£)**					
1 Food & non-alcoholic drinks	**19.60**	**37.70**	**24.40**	**47.00**	**21.30**	**43.50**
2 Alcoholic drinks, tobacco & narcotics	**3.50**	**5.90**	**4.60**	**10.10**	**8.30**	**13.00**
3 Clothing & footwear	**4.90**	**6.50**	**6.40**	**13.90**	**11.00**	**24.10**
4 Housing[2], fuel & power	**19.80**	**24.30**	**29.40**	**29.30**	**32.50**	**39.80**
5 Household goods & services	**9.00**	**14.90**	**18.10**	**27.20**	**19.90**	**36.30**
6 Health	**2.00**	**3.60**	**3.00**	**8.30**	**3.10**	**6.50**
7 Transport	**5.20**	**16.70**	**20.10**	**40.50**	**36.00**	**80.50**
8 Communication	**4.40**	**5.10**	**5.30**	**7.10**	**8.00**	**10.50**
9 Recreation & culture	**13.20**	**25.20**	**22.60**	**51.80**	**35.30**	**66.70**
10 Education	**..**	**..**	**..**	**[2.00]**	**1.80**	**2.90**
11 Restaurants & hotels	**5.10**	**9.30**	**10.10**	**25.00**	**22.70**	**41.80**
12 Miscellaneous goods & services	**8.30**	**13.90**	**17.00**	**29.70**	**20.70**	**38.10**
1-12 All expenditure groups	**95.30**	**162.90**	**161.20**	**292.00**	**220.60**	**403.70**
13 Other expenditure items	**8.30**	**13.80**	**23.60**	**33.60**	**47.70**	**77.10**
Total expenditure	**103.50**	**176.70**	**184.70**	**325.60**	**268.30**	**480.70**
Average weekly expenditure per person (£)						
Total expenditure	**103.50**	**88.40**	**184.70**	**162.80**	**268.30**	**240.40**

Note: The commodity and service categories are not comparable with those in publications before 2001-02
1 Mainly dependent on state pension and not economically active - see appendix D
2 Excluding mortgage interest payments, council tax and Northern Ireland rates

4.1 Expenditure by household composition (cont.) 2002-03

based on weighted data and including children's expenditure

	Retired and non-retired households						
	One adult		Two adults			Three or more adults	
	with one child	with two or more children	with one child	with two children	with three or more children	without children	with children
Grossed number of households (thousands)	700	820	1,840	2,230	900	2,240	880
Total number of households in sample	224	272	536	696	290	477	234
Total number of persons in sample	448	924	1,608	2,784	1,546	1,585	1,148
Total number of adults in sample	224	272	1072	1,392	580	1,585	782
Weighted average number of persons per household	2.0	3.4	3.0	4.0	5.3	3.3	4.9
Commodity or service	**Average weekly household expenditure (£)**						
1 Food & non-alcoholic drinks	**29.30**	**40.00**	**52.90**	**62.30**	**64.50**	**60.70**	**70.10**
2 Alcoholic drinks, tobacco & narcotics	**8.10**	**10.80**	**14.20**	**12.90**	**14.70**	**18.50**	**17.60**
3 Clothing & footwear	**15.70**	**25.20**	**30.80**	**36.20**	**35.00**	**35.20**	**51.30**
4 Housing[2], fuel & power	**30.70**	**31.50**	**40.40**	**39.80**	**44.30**	**53.40**	**45.10**
5 Household goods & services	**16.30**	**18.10**	**46.40**	**37.70**	**43.20**	**42.20**	**39.30**
6 Health	**1.30**	**1.50**	**6.00**	**4.60**	**3.40**	**6.80**	**4.20**
7 Transport	**25.30**	**24.60**	**75.00**	**85.60**	**65.60**	**106.30**	**101.40**
8 Communication	**9.70**	**10.70**	**13.60**	**14.20**	**12.60**	**17.30**	**18.90**
9 Recreation & culture	**31.10**	**36.10**	**74.50**	**81.40**	**85.00**	**89.80**	**80.70**
10 Education	**4.70**	**2.20**	**15.40**	**14.30**	**11.80**	**7.30**	**9.70**
11 Restaurants & hotels	**18.90**	**22.80**	**44.00**	**47.80**	**49.50**	**69.30**	**62.90**
12 Miscellaneous goods & services	**23.50**	**23.00**	**48.70**	**53.60**	**38.20**	**44.80**	**49.60**
1-12 All expenditure groups	**214.70**	**246.60**	**461.90**	**490.40**	**467.90**	**551.80**	**550.70**
13 Other expenditure items	**29.10**	**29.00**	**83.40**	**94.10**	**78.00**	**71.30**	**83.40**
Total expenditure	**243.80**	**275.60**	**545.30**	**584.60**	**545.90**	**623.10**	**634.10**
Average weekly expenditure per person (£)							
Total expenditure	**121.90**	**81.30**	**181.80**	**146.10**	**102.40**	**188.30**	**130.10**

Note: The commodity and service categories are not comparable with those in publications before 2001-02
1 Mainly dependent on state pension and not economically active - see appendix D
2 Excluding mortgage interest payments, council tax and Northern Ireland rates

4.2 Expenditure of one person retired households mainly dependent on state pensions[1] by gross income quintile group

2001-02 – 2002-03

based on weighted data

		Lowest twenty per cent	Second quintile group	Third quintile group	Fourth quintile group	Highest twenty per cent	All house-holds
Lower boundary of group (£ per week)[2]			188	341	541	821	
Average grossed number of households (thousands)		1,520	110	0	0	0	1,630
Total number of households in sample (over 2 years)		869	61	0	0	0	930
Total number of persons in sample (over 2 years)		869	61	0	0	0	930
Total number of adults in sample (over 2 years)		869	61	0	0	0	930
Weighted average number of persons per household		1.0	1.0	0	0	0	1.0
Commodity or service		**Average weekly household expenditure (£)**					
1	**Food & non-alcoholic drinks**	**20.10**	**21.20**	..	..	..	**20.10**
2	**Alcoholic drinks, tobacco & narcotics**	**3.40**	**[3.10]**	..	..	..	**3.30**
3	**Clothing & footwear**	**4.70**	**10.00**	..	..	..	**5.10**
4	**Housing[3], fuel & power**	**19.70**	**18.40**	..	..	..	**19.60**
5	**Household goods & services**	**8.90**	**17.40**	..	..	..	**9.50**
6	**Health**	**1.90**	**1.10**	..	..	..	**1.80**
7	**Transport**	**4.60**	**8.00**	..	..	..	**4.90**
8	**Communication**	**4.30**	**4.60**	..	..	..	**4.30**
9	**Recreation & culture**	**12.10**	**16.00**	..	..	..	**12.30**
10	**Education**	..	..	..	..	..	..
11	**Restaurants & hotels**	**5.10**	**6.60**	..	..	..	**5.20**
12	**Miscellaneous goods & services**	**8.10**	**11.80**	..	..	..	**8.40**
1-12	**All expenditure groups**	**92.90**	**118.20**	..	..	..	**94.60**
13	**Other expenditure items**	**7.80**	**8.50**	..	..	..	**7.80**
Total expenditure		**100.70**	**126.70**	..	..	..	**102.40**
Average weekly expenditure per person (£)							
Total expenditure		**100.70**	**126.70**	..	..	..	**102.40**

Note: The commodity and service categories are not comparable with those in publications before 2001-02
1 Mainly dependent on state pension and not economically active - see appendix D
2 Lower boundary of 2002-03 gross income quintile groups (£ per week)
3 Excluding mortgage interest payments, council tax and Northern Ireland rates

4.3 Expenditure of one person retired households not mainly dependent on state pensions by gross income quintile group 2001-02 – 2002-03

based on weighted data

		Lowest twenty per cent	Second quintile group	Third quintile group	Fourth quintile group	Highest twenty per cent	All house-holds
Lower boundary of group (£ per week)[1]			188	341	541	821	
Average grossed number of households (thousands)		600	670	170	60	30	1,530
Total number of households in sample (over 2 years)		352	381	88	33	15	869
Total number of persons in sample (over 2 years)		352	381	88	33	15	869
Total number of adults in sample (over 2 years)		352	381	88	33	15	869
Weighted average number of persons per household		1.0	1.0	1.0	1.0	1.0	1.0
Commodity or service		**Average weekly household expenditure (£)**					
1	Food & non-alcoholic drinks	20.60	24.80	25.60	33.00	[35.90]	23.80
2	Alcoholic drinks, tobacco & narcotics	3.10	5.20	4.90	[6.20]	..	4.40
3	Clothing & footwear	4.70	6.70	11.20	[19.00]	..	7.10
4	Housing[2], fuel & power	26.60	28.00	29.90	32.50	[75.00]	28.80
5	Household goods & services	13.60	19.90	20.80	33.10	[21.50]	18.00
6	Health	1.70	3.10	7.10	[1.40]	..	3.10
7	Transport	10.40	19.80	30.40	52.50	[61.10]	19.40
8	Communication	4.50	5.40	5.60	8.50	[10.30]	5.30
9	Recreation & culture	16.50	23.70	28.90	55.10	[64.50]	23.60
10	Education	..	..	..	..	..	..
11	Restaurants & hotels	7.60	9.40	14.80	13.50	..	9.80
12	Miscellaneous goods & services	10.40	17.90	27.10	36.60	[142.50]	19.10
1-12	All expenditure groups	119.80	164.20	206.30	295.70	[480.10]	162.90
13	Other expenditure items	17.80	24.00	28.10	39.10	[144.30]	25.10
Total expenditure		137.60	188.30	234.50	334.90	[624.40]	188.00
Average weekly expenditure per person (£)							
Total expenditure		137.60	188.30	234.50	334.90	[624.40]	188.00

Note: The commodity and service categories are not comparable with those in publications before 2001-02
1 Lower boundary of 2002-03 gross income quintile groups (£ per week)
2 Excluding mortgage interest payments, council tax and Northern Ireland rates

4.4 Expenditure of one person non-retired households by gross income quintile group 2001-02 – 2002-03

based on weighted data

	Lowest twenty per cent	Second quintile group	Third quintile group	Fourth quintile group	Highest twenty per cent	All households
Lower boundary of group (£ per week)[1]		188	341	541	821	
Average grossed number of households (thousands)	1,100	950	870	440	210	3,560
Total number of households in sample (over 2 years)	667	567	494	234	114	2,076
Total number of persons in sample (over 2 years)	667	567	494	234	114	2,076
Total number of adults in sample (over 2 years)	667	567	494	234	114	2,076
Weighted average number of persons per household	1.0	1.0	1.0	1.0	1.0	1.0
Commodity or service		**Average weekly household expenditure (£)**				
1 Food & non-alcoholic drinks	**17.80**	**21.10**	**22.60**	**23.80**	**26.00**	**21.10**
2 Alcoholic drinks, tobacco & narcotics	**7.70**	**8.10**	**7.80**	**10.00**	**11.10**	**8.30**
3 Clothing & footwear	**5.30**	**9.90**	**11.90**	**14.80**	**32.50**	**10.90**
4 Housing[2], fuel & power	**24.90**	**33.10**	**34.40**	**36.80**	**44.50**	**32.00**
5 Household goods & services	**11.10**	**16.40**	**22.80**	**24.80**	**43.20**	**18.90**
6 Health	**1.30**	**2.60**	**3.00**	**4.40**	**4.50**	**2.60**
7 Transport	**14.60**	**27.60**	**45.10**	**79.40**	**69.70**	**36.70**
8 Communication	**5.50**	**7.10**	**9.70**	**10.10**	**13.40**	**8.00**
9 Recreation & culture	**17.50**	**25.50**	**36.10**	**63.40**	**77.60**	**33.40**
10 Education	**[1.00]**	**[0.60]**	**2.00**	**[2.60]**	**..**	**1.80**
11 Restaurants & hotels	**11.50**	**16.70**	**25.80**	**37.30**	**53.60**	**22.00**
12 Miscellaneous goods & services	**8.90**	**16.30**	**24.50**	**26.40**	**48.70**	**19.10**
1-12 All expenditure groups	**127.10**	**185.10**	**245.90**	**333.90**	**433.60**	**214.70**
13 Other expenditure items	**13.20**	**35.80**	**61.20**	**82.10**	**143.20**	**46.90**
Total expenditure	**140.40**	**220.80**	**307.10**	**416.00**	**576.80**	**261.60**
Average weekly expenditure per person (£)						
Total expenditure	**140.40**	**220.80**	**307.10**	**416.00**	**576.80**	**261.60**

Note: The commodity and service categories are not comparable with those in publications before 2001-02
1 Lower boundary of 2002-03 gross income quintile groups (£ per week)
2 Excluding mortgage interest payments, council tax and Northern Ireland rates

4.5 Expenditure of one adult households with children by gross income quintile group 2001-02 – 2002-03

based on weighted data and including children's expenditure

	Lowest twenty per cent	Second quintile group	Third quintile group	Fourth quintile group	Highest twenty per cent	All households
Lower boundary of group (£ per week)[1]		188	341	541	821	
Average grossed number of households (thousands)	650	390	310	130	20	1,500
Total number of households in sample (over 2 years)	446	273	221	79	15	1,034
Total number of persons in sample (over 2 years)	1,170	836	601	213	41	2,861
Total number of adults in sample (over 2 years)	446	273	221	79	15	1,034
Weighted average number of persons per household	2.6	2.9	2.8	2.7	2.9	2.7
Commodity or service	**Average weekly household expenditure (£)**					
1 Food & non-alcoholic drinks	**30.30**	**35.40**	**36.70**	**48.30**	**[51.80]**	**34.70**
2 Alcoholic drinks, tobacco & narcotics	**7.90**	**8.60**	**9.50**	**12.60**	**[12.30]**	**8.90**
3 Clothing & footwear	**14.80**	**21.60**	**27.80**	**26.60**	**[59.20]**	**21.10**
4 Housing[2], fuel & power	**21.50**	**34.00**	**41.90**	**40.10**	**[27.40]**	**30.80**
5 Household goods & services	**13.60**	**16.20**	**24.00**	**32.00**	**[61.50]**	**18.50**
6 Health	**0.70**	**1.60**	**2.00**	**7.80**	**..**	**1.90**
7 Transport	**12.90**	**24.00**	**38.50**	**60.30**	**[89.90]**	**26.40**
8 Communication	**7.10**	**11.30**	**13.10**	**13.00**	**[18.70]**	**10.20**
9 Recreation & culture	**20.70**	**30.90**	**46.00**	**71.10**	**[98.20]**	**34.10**
10 Education	**1.50**	**3.70**	**5.40**	**19.40**	**..**	**5.10**
11 Restaurants & hotels	**13.60**	**19.10**	**28.80**	**36.40**	**[38.90]**	**20.40**
12 Miscellaneous goods & services	**11.20**	**20.90**	**31.00**	**52.30**	**[77.10]**	**22.30**
1-12 All expenditure groups	**155.80**	**227.30**	**304.60**	**419.90**	**[579.40]**	**234.50**
13 Other expenditure items	**9.30**	**24.80**	**43.20**	**68.20**	**[150.00]**	**27.60**
Total expenditure	**165.10**	**252.10**	**347.80**	**488.10**	**[729.30]**	**262.10**
Average weekly expenditure per person (£) Total expenditure	**63.10**	**86.70**	**123.70**	**182.40**	**[248.50]**	**95.40**

Note: The commodity and service categories are not comparable with those in publications before 2001-02

1 Lower boundary of 2002-03 gross income quintile groups (£ per week)

2 Excluding mortgage interest payments, council tax and Northern Ireland rates

4.6 Expenditure of two adult households with children by gross income quintile group

2001-02 – 2002-03

based on weighted data and including children's expenditure

	Lowest twenty per cent	Second quintile group	Third quintile group	Fourth quintile group	Highest twenty per cent	All house-holds
Lower boundary of group (£ per week)[1]		188	341	541	821	
Average grossed number of households (thousands)	190	550	1,140	1,550	1,530	4,950
Total number of households in sample (over 2 years)	120	368	757	1,005	950	3,200
Total number of persons in sample (over 2 years)	453	1,483	2,995	3,871	3,693	12,495
Total number of adults in sample (over 2 years)	240	736	1,514	2,010	1,900	6,400
Weighted average number of persons per household	3.6	4.0	3.9	3.8	3.8	3.9
Commodity or service		**Average weekly household expenditure (£)**				
1 Food & non-alcoholic drinks	**44.20**	**47.90**	**51.50**	**58.20**	**71.00**	**58.90**
2 Alcoholic drinks, tobacco & narcotics	**13.90**	**13.30**	**13.70**	**13.00**	**14.30**	**13.70**
3 Clothing & footwear	**21.20**	**22.80**	**25.80**	**33.60**	**46.70**	**34.20**
4 Housing[2], fuel & power	**27.70**	**36.20**	**37.70**	**35.80**	**48.10**	**39.80**
5 Household goods & services	**22.70**	**26.50**	**28.10**	**37.90**	**64.80**	**42.10**
6 Health	**1.70**	**1.50**	**2.60**	**4.60**	**6.90**	**4.40**
7 Transport	**40.40**	**41.40**	**54.70**	**71.00**	**118.80**	**77.50**
8 Communication	**9.30**	**11.80**	**11.40**	**12.40**	**15.80**	**13.00**
9 Recreation & culture	**32.10**	**48.50**	**54.90**	**74.50**	**113.40**	**77.50**
10 Education	**[4.00]**	**1.60**	**5.00**	**7.00**	**36.90**	**15.10**
11 Restaurants & hotels	**18.70**	**27.40**	**31.30**	**44.00**	**64.00**	**44.40**
12 Miscellaneous goods & services	**20.20**	**21.30**	**30.20**	**42.50**	**76.70**	**47.00**
1-12 All expenditure groups	**256.00**	**300.20**	**346.90**	**434.70**	**677.50**	**467.60**
13 Other expenditure items	**28.30**	**36.10**	**55.30**	**83.80**	**147.20**	**89.30**
Total expenditure	**284.30**	**336.30**	**402.20**	**518.50**	**824.60**	**556.90**
Average weekly expenditure per person (£)						
Total expenditure	**79.00**	**84.90**	**102.40**	**134.80**	**214.20**	**143.80**

Note: The commodity and service categories are not comparable with those in publications before 2001-02
1 Lower boundary of 2002-03 gross income quintile groups (£ per week)
2 Excluding mortgage interest payments, council tax and Northern Ireland rates

4.7 Expenditure of one man one woman non-retired households by gross income quintile group — 2001-02 – 2002-03

based on weighted data

	Lowest twenty per cent	Second quintile group	Third quintile group	Fourth quintile group	Highest twenty per cent	All house-holds
Lower boundary of group (£ per week)[1]		188	341	541	821	
Average grossed number of households (thousands)	300	630	1190	1460	1420	5,010
Total number of households in sample (over 2 years)	188	395	705	839	771	2,898
Total number of persons in sample (over 2 years)	376	790	1,410	1,678	1,542	5,796
Total number of adults in sample (over 2 years)	376	790	1,410	1,678	1,542	5,796
Weighted average number of persons per household	2.0	2.0	2.0	2.0	2.0	2.0
Commodity or service		**Average weekly household expenditure (£)**				
1 Food & non-alcoholic drinks	36.10	39.10	40.70	42.60	48.00	42.80
2 Alcoholic drinks, tobacco & narcotics	10.90	12.40	13.10	13.30	15.50	13.60
3 Clothing & footwear	11.10	11.80	16.80	24.50	37.90	24.10
4 Housing[2], fuel & power	31.90	32.50	38.20	35.70	47.80	39.10
5 Household goods & services	23.80	30.60	24.90	39.20	55.40	38.40
6 Health	2.00	4.50	5.20	6.70	11.60	7.20
7 Transport	40.50	36.00	60.50	78.40	123.20	79.20
8 Communication	6.30	7.80	8.90	10.40	13.40	10.30
9 Recreation & culture	35.10	40.30	53.20	65.10	94.00	65.50
10 Education	..	[0.70]	1.80	1.40	5.40	2.60
11 Restaurants & hotels	19.60	22.10	28.40	39.10	66.50	41.00
12 Miscellaneous goods & services	21.90	20.00	28.60	37.60	52.60	36.60
1-12 All expenditure groups	241.70	257.70	320.30	394.20	571.40	400.50
13 Other expenditure items	24.60	35.80	52.00	71.40	135.80	77.80
Total expenditure	266.40	293.60	372.30	465.50	707.20	478.30
Average weekly expenditure per person (£) Total expenditure	133.20	146.80	186.20	232.80	353.60	239.20

Note: The commodity and service categories are not comparable with those in publications before 2001-02

1 Lower boundary of 2002-03 gross income quintile groups (£ per week)

2 Excluding mortgage interest payments, council tax and Northern Ireland rates

4.8 Expenditure of one man one woman retired households mainly dependent on state pensions[1] by gross income quintile group 2001-02 – 2002-03

based on weighted data

	Lowest twenty per cent	Second quintile group	Third quintile group	Fourth quintile group	Highest twenty per cent	All house-holds
Lower boundary of group (£ per week)[2]		188	341	541	821	
Average grossed number of households (thousands)	350	340	10	0	0	699
Total number of households in sample (over 2 years)	215	212	5	0	0	432
Total number of persons in sample (over 2 years)	430	424	10	0	0	864
Total number of adults in sample (over 2 years)	430	424	10	0	0	864
Weighted average number of persons per household	2.0	2.0	2.0	0	0	2.0
Commodity or service		Average weekly household expenditure (£)				
1 Food & non-alcoholic drinks	36.30	38.30	..	..	..	37.30
2 Alcoholic drinks, tobacco & narcotics	5.40	7.00	..	..	..	6.10
3 Clothing & footwear	6.30	8.40	..	..	..	7.40
4 Housing[3], fuel & power	24.70	26.90	..	..	..	25.60
5 Household goods & services	10.20	17.20	..	..	..	14.70
6 Health	2.90	4.30	..	..	..	3.70
7 Transport	16.20	22.10	..	..	..	19.00
8 Communication	4.80	4.90	..	..	..	4.90
9 Recreation & culture	25.20	27.50	..	..	..	26.50
10 Education	..	..	..	..	..	..
11 Restaurants & hotels	8.50	11.70	..	..	..	10.20
12 Miscellaneous goods & services	13.40	15.70	..	..	..	14.60
1-12 All expenditure groups	153.80	184.20	..	..	..	169.90
13 Other expenditure items	14.90	17.00	..	..	..	15.80
Total expenditure	168.70	201.10	..	..	..	185.80
Average weekly expenditure per person (£)						
Total expenditure	84.40	100.60	..	..	..	92.90

Note: The commodity and service categories are not comparable with those in publications before 2001-02

1 Mainly dependent on state pension and not economically active - see appendix D

2 Lower boundary of 2002-03 gross income quintile groups (£ per week)

3 Excluding mortgage interest payments, council tax and Northern Ireland rates

4.9 Expenditure of one man one woman retired households not mainly dependent on state pensions by gross income quintile group

2001-02 – 2002-03

based on weighted data

	Lowest twenty per cent	Second quintile group	Third quintile group	Fourth quintile group	Highest twenty per cent	All house-holds
Lower boundary of group (£ per week)[1]		188	341	541	821	
Average grossed number of households (thousands)	50	860	610	240	100	1,869
Total number of households in sample (over 2 years)	31	523	376	139	62	1,131
Total number of persons in sample (over 2 years)	62	1,046	752	278	124	2,262
Total number of adults in sample (over 2 years)	62	1,046	752	278	124	2,262
Weighted average number of persons per household	2.0	2.0	2.0	2.0	2.0	2.0
Commodity or service	**Average weekly household expenditure (£)**					
1 Food & non-alcoholic drinks	**38.50**	**41.50**	**45.30**	**51.30**	**64.00**	**45.10**
2 Alcoholic drinks, tobacco & narcotics	**[8.20]**	**7.60**	**8.80**	**12.10**	**14.70**	**9.00**
3 Clothing & footwear	**7.70**	**11.50**	**18.20**	**17.40**	**28.70**	**15.00**
4 Housing[2], fuel & power	**27.60**	**28.00**	**27.80**	**36.70**	**40.30**	**29.80**
5 Household goods & services	**20.90**	**21.00**	**27.90**	**38.80**	**45.40**	**26.90**
6 Health	**[1.00]**	**5.50**	**9.90**	**8.90**	**7.00**	**7.30**
7 Transport	**15.00**	**32.20**	**48.10**	**49.20**	**76.20**	**41.10**
8 Communication	**5.70**	**5.80**	**6.60**	**8.10**	**10.90**	**6.60**
9 Recreation & culture	**23.90**	**40.50**	**49.80**	**64.40**	**92.90**	**48.90**
10 Education	**..**	**..**	**..**	**..**	**..**	**[1.20]**
11 Restaurants & hotels	**11.70**	**16.30**	**25.10**	**37.70**	**39.40**	**22.90**
12 Miscellaneous goods & services	**15.20**	**17.90**	**29.30**	**42.40**	**64.60**	**27.20**
1-12 All expenditure groups	**176.80**	**227.90**	**296.90**	**368.20**	**497.70**	**281.10**
13 Other expenditure items	**15.10**	**24.20**	**33.30**	**43.50**	**83.90**	**32.50**
Total expenditure	**191.90**	**252.10**	**330.20**	**411.70**	**581.70**	**313.50**
Average weekly expenditure per person (£)						
Total expenditure	**95.90**	**126.00**	**165.10**	**205.80**	**290.80**	**156.80**

Note: The commodity and service categories are not comparable with those in publications before 2001-02

1 Lower boundary of 2002-03 gross income quintile groups (£ per week)

2 Excluding mortgage interest payments, council tax and Northern Ireland rates

4.10 Household expenditure by tenure

based on weighted data

2002-03

		Owners			Social rented from		
		Owned outright	Buying with a mortgage [1]	All	Council [2]	Registered Social Landlord [3]	All
Grossed number of households (thousands)		7,090	10,030	17,130	3,420	1,440	4,850
Total number of households in sample		2,047	2,836	4,883	1,012	400	1,412
Total number of persons in sample		3,951	8,026	11,977	2,347	863	3,210
Total number of adults in sample		3,678	5,561	9,239	1,569	573	2,142
Weighted average number of persons per household		1.9	2.8	2.4	2.3	2.1	2.3
Commodity or service		**Average weekly household expenditure (£)**					
1	**Food & non-alcoholic drinks**	**43.10**	**49.60**	**46.90**	**32.40**	**31.60**	**32.10**
2	**Alcoholic drinks, tobacco & narcotics**	**9.20**	**13.20**	**11.60**	**11.90**	**8.50**	**10.90**
3	**Clothing & footwear**	**17.40**	**30.70**	**25.20**	**13.20**	**14.20**	**13.50**
4	**Housing[6], fuel & power**	**28.40**	**29.50**	**29.10**	**36.60**	**41.20**	**37.90**
5	**Household goods & services**	**29.10**	**40.70**	**35.90**	**16.00**	**14.80**	**15.70**
6	**Health**	**6.20**	**5.80**	**6.00**	**1.60**	**1.70**	**1.60**
7	**Transport**	**54.70**	**82.90**	**71.20**	**22.90**	**22.80**	**22.90**
8	**Communication**	**8.20**	**13.00**	**11.00**	**7.90**	**8.70**	**8.20**
9	**Recreation & culture**	**53.70**	**74.50**	**65.90**	**28.60**	**28.10**	**28.50**
10	**Education**	**3.20**	**8.40**	**6.20**	**0.90**	**0.50**	**0.80**
11	**Restaurants & hotels**	**28.20**	**48.20**	**39.90**	**17.80**	**16.00**	**17.30**
12	**Miscellaneous goods & services**	**32.10**	**45.40**	**39.90**	**13.30**	**13.10**	**13.30**
1-12	**All expenditure groups**	**313.60**	**441.90**	**388.80**	**203.30**	**201.30**	**202.70**
13	**Other expenditure items**	**33.80**	**103.00**	**74.30**	**13.00**	**13.20**	**13.10**
Total expenditure		**347.40**	**544.90**	**463.10**	**216.30**	**214.50**	**215.80**
Average weekly expenditure per person (£)							
Total expenditure		**179.20**	**196.10**	**190.50**	**92.80**	**101.20**	**95.10**

Note: The commodity and service categories are not comparable with those in publications before 2001-02
1 Including shared owners (who own part of the equity and pay mortgage, part rent).
2 "Council" includes local authorities, New Towns and Scottish Homes, but see note 3 below.
3 Formerly Housing Associations

4.10 Household expenditure by tenure (cont.)

2002-03

based on weighted data

	Private rented [4]				All tenures
	Rent free	Rent paid, unfurnished [5]	Rent paid, furnished	All	
Grossed number of households (thousands)	350	1,440	580	2,360	24,350
Total number of households in sample	100	393	139	632	6,927
Total number of persons in sample	191	895	313	1,399	16,586
Total number of adults in sample	153	630	286	1,069	12,450
Weighted average number of persons per household	2.0	2.3	2.4	2.3	2.4
Commodity or service	**Average weekly household expenditure (£)**				
1 Food & non-alcoholic drinks	**31.00**	**35.30**	**33.00**	**34.10**	**42.70**
2 Alcoholic drinks, tobacco & narcotics	**8.30**	**11.80**	**10.50**	**11.00**	**11.40**
3 Clothing & footwear	**14.70**	**18.70**	**23.90**	**19.40**	**22.30**
4 Housing[6], fuel & power	**15.10**	**91.30**	**137.50**	**91.40**	**36.90**
5 Household goods & services	**12.80**	**24.60**	**8.30**	**18.90**	**30.20**
6 Health	**2.20**	**2.90**	**3.70**	**3.00**	**4.80**
7 Transport	**34.50**	**44.50**	**60.40**	**46.90**	**59.20**
8 Communication	**8.60**	**12.70**	**15.30**	**12.70**	**10.60**
9 Recreation & culture	**37.50**	**40.30**	**60.90**	**44.90**	**56.40**
10 Education	..	**4.50**	**15.80**	**6.70**	**5.20**
11 Restaurants & hotels	**24.70**	**36.50**	**58.60**	**40.20**	**35.40**
12 Miscellaneous goods & services	**21.10**	**26.50**	**21.80**	**24.50**	**33.10**
1-12 All expenditure groups	**210.80**	**349.60**	**449.80**	**353.70**	**348.30**
13 Other expenditure items	**20.50**	**30.80**	**37.50**	**30.90**	**57.90**
Total expenditure	**231.40**	**380.40**	**487.30**	**384.60**	**406.20**
Average weekly expenditure per person (£)					
Total expenditure	**115.20**	**167.90**	**204.20**	**170.40**	**170.50**

Note: The commodity and service categories are not comparable with those in publications before 2001-02

4 All tenants whose accommodation goes with the job of someone in the household are allocated to "rented privately", even if the landlord is a local authority or housing association or Housing Action Trust, or if the accommodation is rent free. Squatters are also included in this category.

5 "Unfurnished" includes the answers: "partly furnished".

6 Excluding mortgage interest payments, council tax and Northern Ireland rates

Statistical Regions of the United Kingdom

Chapter 5

Expenditure by region

- Averaged over the last two years total expenditure varied from £486 a week in London to £336 in Wales. London, the South East and East of England were the only regions in which average expenditure was higher than the UK average. Spending in the North East, Yorkshire and the Humber, and Wales was between 12 and 16 per cent lower than the UK average.

- Households in Northern Ireland spent a higher proportion on **clothing and footwear** than anywhere else, 8 per cent compared with the UK average of 6 per cent.

- Households in London spent by far the most on **transport services** at £17.90 a week, more than double the UK average of £8.40. However, they spent the least on **petrol, diesel and other motor oils** at £11.00 a week, compared to a UK average of £14.70.

Urban and rural areas

Classification based on the population of the continuous built-up areas, irrespective of administrative boundaries.

- Averaged over the last two years total expenditure was highest in the London built-up area at £487 a week. It was lowest in other metropolitan built-up areas at £336 a week. The highest value outside London was £458, in rural areas.

- **Transport** was the largest item of expenditure for most areas. In other metropolitan built-up areas, other urban areas with a population between 25,000 and 100,000 and urban areas with a population over 250,000, however, spending was highest for **recreation and culture**.

- Rural areas spent more than other areas on **household goods and services**, at £36.90 a week. They were also the highest spending households on **food and non-alcoholic drinks** (£46.30 a week), **transport** (£73.10 a week) **and recreation and culture** (£63.50 a week).

5.1 Total weekly expenditure in relation to the UK average

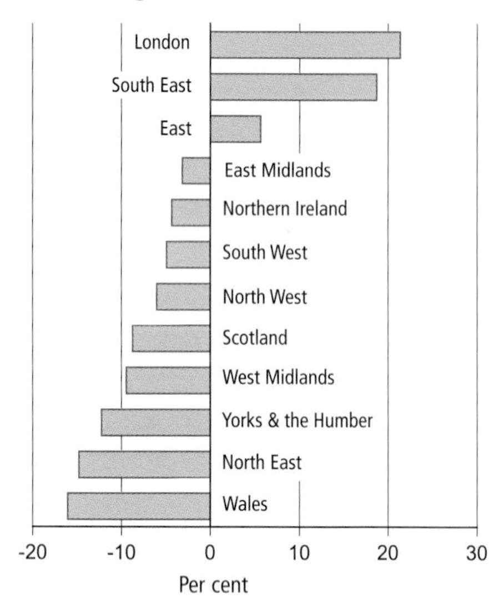

Chapter 5
Expenditure by region

This chapter illustrates the differences in household expenditure patterns between the regions of the United Kingdom and between types of area.

The regional breakdowns are based on Government Office Regions (GORs) in England, and Wales, Scotland and Northern Ireland, a map of which can be found on page 82. Another table shows a breakdown between urban and rural areas.

The tables and analysis in this chapter are based on data from the first two-years of the EFS, 2001-02 – 2002-03. This increases sample size and reliability.

Overview of national and regional differences

Both the level and the composition of expenditure vary across the countries and regions of the UK.

Table 5.1 shows the average weekly expenditure of households by UK countries and Government Office Regions. **Figure 5.1** compares the total average weekly expenditure in each region with that for the UK. Over the last two years, household expenditure in the UK averaged £400 a week. Weekly expenditure for households in London, the South East and the East of England was more than the UK average. Households in London spent an average of £486 a week, around 20 per cent higher than the UK average. Spending in the North East, Yorkshire and the Humber, and Wales was between 12 and 16 per cent lower than the UK average. Wales had the lowest average weekly expenditure at £336.

Table 5.2 shows average weekly household expenditure on the main commodity headings as a percentage of total expenditure. In general, patterns of expenditure were very similar across the regions. Households in Northern Ireland spent the most on clothing and footwear both in monetary terms (£30.20 a week) and as a proportion of total expenditure, 8 per cent compared to a UK average of 6 per cent. While households in London spent the lowest proportion on transport at 13 per cent, it was still one of only three regions in which spending was over £60 a week on this item. The other two were East of England and the South East. Households in each region spent between two and four per cent of total expenditure on alcoholic drink and tobacco. This ranged from households in Scotland and Northern Ireland spending around £13.50 a week (four per cent) to those in the East of England and London spending around £10.00 (two per cent).

Detailed expenditure patterns

Table 5.3 shows detailed household expenditure on the full range of commodities and services by UK countries and GORs.

Meat (beef, pork, lamb, poultry, bacon and ham)

Figure 5.2 shows spending on uncooked meat by region. Northern Ireland spent by far the largest amount each week on fresh, frozen and chilled meat at £6.50, 33 per cent more than the UK average of £4.90. Households across the regions generally spent more on poultry than on other meat types. The UK average for poultry was £1.50 a week compared to £1.30 a week for beef. However, Northern Ireland households spent twice the national average on beef each week, at £2.60 a week. Spending on pork was consistent across the regions at around 50 to 60 pence a week.

The category 'other meat and meat preparations' includes spending on sausages, offal, preserved and processed meat (including meat ready meals) and other meat eg rabbit, venison etc. Households spent nearly as much on these products as on uncooked meats. Households in Scotland and Northern Ireland recorded the highest average spend on these items at over £5.00 a week.

Clothing and footwear

Table 5.3 and **figure 5.3** show that spending on women's outerwear was greater than that for any other item in this category. Households in the South East, Northern Ireland and London spent £9.50 or more a week on women's outerwear compared to a UK average of £8.00 a week. Households in Yorkshire and Humberside spent less than any other region on women's outerwear (£6.20 a week) while those in Wales spent the least on men's outerwear (£3.00 a week). Households in Northern Ireland spent 40 per cent more than the national average on men's outerwear and over 50 per cent more on footwear. Households in the South West spent the least on footwear at £3.00 a week; this compares to a national average of £4.10.

5.2 Expenditure on uncooked meat

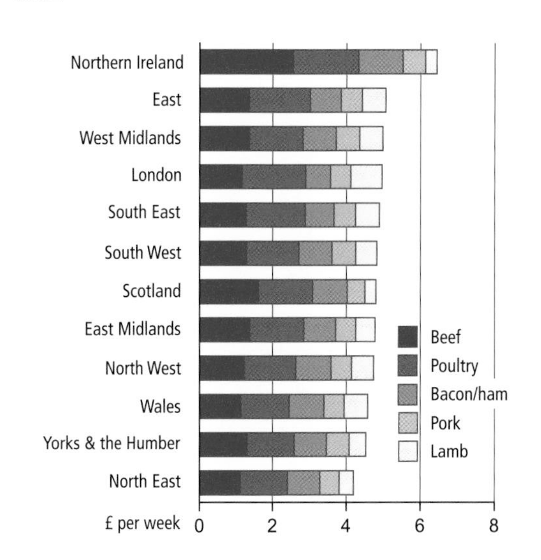

5.3 Expenditure on women's & men's outerwear & footwear

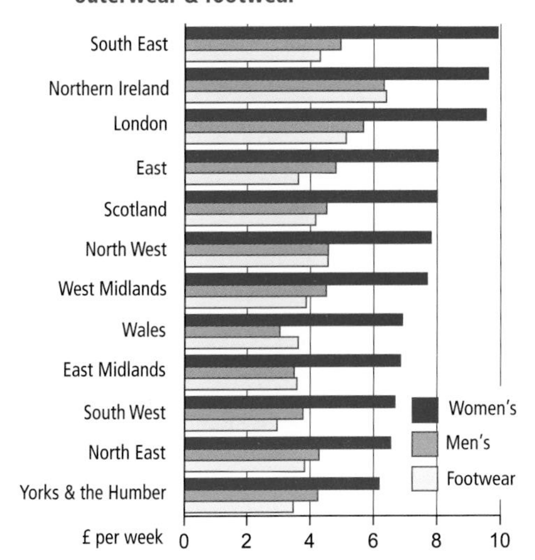

5.4 Expenditure on new and used cars/vans

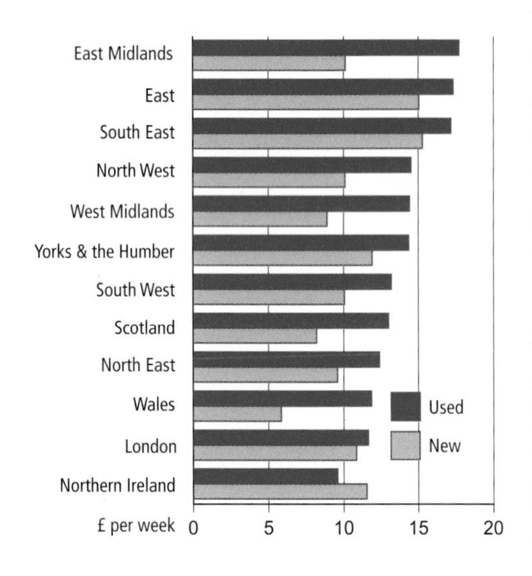

Transport

This category includes the purchase of vehicles, operation of personal transport and transport services (fares and other travel costs). It excludes spending on vehicle insurance, road tax and motoring fines. Spending on new cars in the South East and East of England, at £15.00 or more a week, was around 40 per cent above the UK average of £11.00. Households in Wales recorded the lowest spend on new cars at £5.80 a week, nearly 50 per cent below the UK average. Generally, much more was spent on the purchase of second hand cars than new cars (**figure 5.4**). The average weekly spend on second hand cars across the UK was £14.30. This ranged from over £17.00 in the East Midlands, East of England and South East to £9.50 in Northern Ireland. Households in London spent less than the UK average on buying and running their cars, but by far the most on transport services. At £17.90 a week this expenditure was more than double the UK average of £8.40.

Recreation and culture

This was the second largest area of expenditure in the UK after transport. Households in the South East had the highest spend at £66.50 followed by East of England at £57.30. Households in Northern Ireland spent £48.00, around 13 per cent less than the UK average.

Figure 5.5 compares expenditure on sports admissions, subscriptions and leisure class fees with TV/ satellite services. The UK's average spend on sports admissions etc was £5.20 a week. Only households in London, the South East and East of England spent more than this each week at £7.90, £6.10 and £5.60 a week respectively. Households in Wales spent the least at £3.50 a week, around a third less than the national average. TV/satellite services include expenditure on TV, video and satellite rentals; cable and internet subscriptions and TV licences. Expenditure ranged from £4.10 a week in Northern Ireland, 15 per cent less than the UK average, to £5.20 a week in the North West, around eight per cent above the UK average.

5.5 Expenditure on sports equipment, subscriptions & leisure class fees and TV/ satellite services

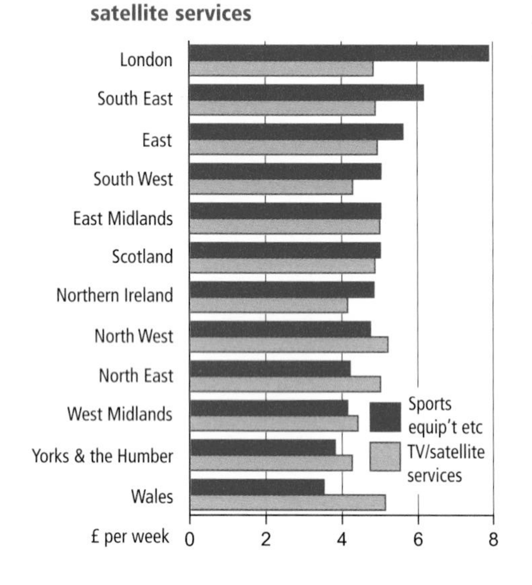

Households in the South East, East of England and the South West spent more on their pets each week than any of the other regions, between £3.60 and £3.90 a week. This included the cost of pets themselves, accessories, pet food and veterinary services. Spending in other regions on this item ranged from £1.70 a week in the North East to £2.80 in the North West.

Type of area

Table 5.4 compares expenditure between urban and rural areas. The classification was developed by the Department for Transport, Local Government and the Regions (DTLR) based on the 1991 Census. The metropolitan built-up areas exclude any rural areas within the metropolitan districts and include any built up areas adjoining them.

Figure 5.6 is a summary of how total spending varied by type of area. Averaged over the last two years, total expenditure was highest in the London built-up area at £487 a week. It was lowest in other metropolitan built-up areas, at £336 a week. Outside London, expenditure was highest in rural areas at £458 a week.

Transport was the largest item of expenditure for most areas. **Figure 5.7** shows that households in rural areas spent by far the most on getting around, at £73.10 a week. This was around £10 a week more than households in the London built up area and £30 more than in other metropolitan built-up areas.

Households in rural areas were the highest spenders on food and non-alcoholic drinks, household goods and services and recreation and culture. They were also the second largest spenders for a number of other items - education, housing, fuel and power, health and communication. Spending on education was highest in the London built-up area at £10.00 a week, followed by £8.20 a week in rural areas. All the other areas spent significantly less on education.

5.6 Expenditure by urban/rural areas

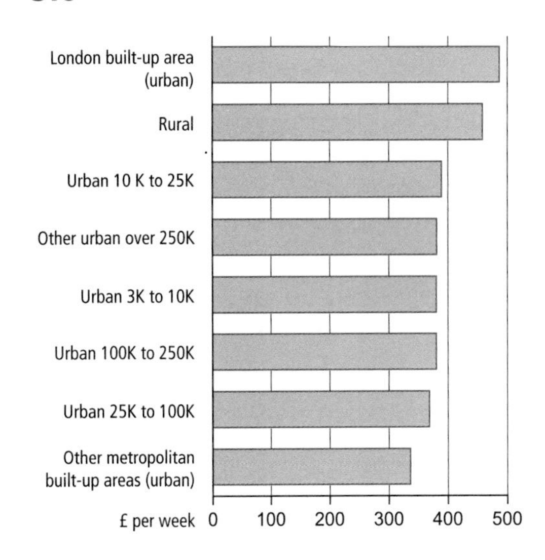

5.7 Expenditure on transport by urban/rural areas

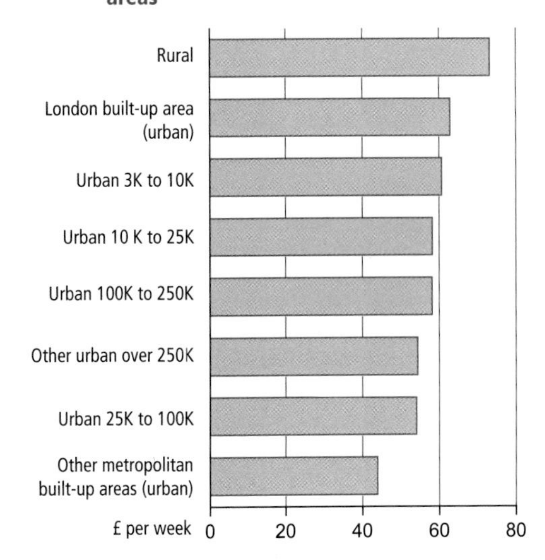

5.1 Household expenditure by UK Countries and Government Office Regions

2001-02 – 2002-03

based on weighted data and including children's expenditure

		North East	North West	Yorks & the Humber	East Midlands	West Midlands	East	London
Average grossed number of households (thousands)		1,020	2,780	2,130	1,720	2,120	2,230	2,850
Total number of households in sample (over 2 years)		632	1,599	1,159	974	1,203	1,278	1,283
Total number of persons in sample (over 2 years)		1,565	3,872	2,698	2,353	3,022	3,075	3,135
Total number of adults in sample (over 2 years)		1,142	2,877	2,029	1,767	2,189	2,313	2,287
Weighted average number of persons per household		2.4	2.4	2.3	2.4	2.5	2.4	2.5
Commodity or service		**Average weekly household expenditure (£)**						
1	**Food & non-alcoholic drinks**	**38.90**	**40.40**	**37.40**	**42.60**	**40.90**	**45.10**	**44.50**
2	**Alcoholic drinks, tobacco & narcotics**	**10.90**	**12.10**	**11.00**	**11.70**	**11.60**	**9.90**	**10.30**
3	**Clothing & footwear**	**20.50**	**23.20**	**18.80**	**19.40**	**22.00**	**22.60**	**27.50**
4	**Housing[1], fuel & power**	**29.30**	**32.10**	**31.40**	**31.40**	**31.20**	**37.40**	**55.50**
5	**Household goods & services**	**23.60**	**29.20**	**26.90**	**30.40**	**27.10**	**30.40**	**33.80**
6	**Health**	**3.20**	**3.40**	**3.80**	**3.90**	**3.50**	**6.80**	**6.40**
7	**Transport**	**49.60**	**53.70**	**52.20**	**59.10**	**53.10**	**69.40**	**62.20**
8	**Communication**	**9.80**	**9.30**	**9.00**	**9.60**	**9.70**	**10.90**	**14.50**
9	**Recreation & culture**	**50.20**	**55.40**	**51.40**	**56.50**	**51.60**	**57.30**	**56.70**
10	**Education**	**3.40**	**5.20**	**3.70**	**3.80**	**2.80**	**5.30**	**10.30**
11	**Restaurants & hotels**	**32.00**	**32.30**	**33.80**	**34.50**	**31.40**	**33.60**	**43.80**
12	**Miscellaneous goods & services**	**24.60**	**31.50**	**25.50**	**30.20**	**28.10**	**34.30**	**39.40**
1-12	**All expenditure groups**	**296.10**	**327.60**	**305.00**	**333.00**	**312.90**	**362.90**	**404.80**
13	**Other expenditure items**	**44.80**	**48.40**	**46.00**	**54.50**	**49.30**	**59.70**	**80.70**
Total expenditure		**340.80**	**376.00**	**351.00**	**387.50**	**362.20**	**422.60**	**485.60**
Average weekly expenditure per person (£)								
Total expenditure		**141.00**	**157.90**	**152.60**	**161.50**	**147.60**	**177.50**	**194.00**

Note: The commodity and service categories are not comparable with those in publications before 2001-02
1 Excluding mortgage interest payments, council tax and Northern Ireland rates

5.1 Household expenditure by UK Countries and Government Office Regions (cont.)

2001-02 – 2002-03

based on weighted data and including children's expenditure

	South East	South West	England	Wales	Scotland	Northern Ireland	United Kingdom
Grossed number of households (thousands)	3,390	2,160	20,390	1,250	2,140	620	24,400
Total number of households in sample (over 2 years)	1,955	1,282	11,365	712	1,207	1,116	14,400
Total number of persons in sample (over 2 years)	4,746	2,925	27,391	1,644	2,759	2,914	34,708
Total number of adults in sample (over 2 years)	3,584	2,275	20,463	1,233	2,112	2,088	25,896
Weighted average number of persons per household	2.3	2.2	2.4	2.3	2.3	2.7	2.4
Commodity or service		**Average weekly household expenditure (£)**					
1 Food & non-alcoholic drinks	44.90	42.20	42.30	39.60	42.20	46.20	42.20
2 Alcoholic drinks, tobacco & narcotics	11.10	10.80	11.00	11.20	13.40	13.50	11.30
3 Clothing & footwear	25.40	18.00	22.50	17.80	23.10	30.20	22.50
4 Housing[1], fuel & power	40.20	36.10	37.30	31.90	32.40	30.30	36.50
5 Household goods & services	35.40	30.60	30.50	26.20	26.60	31.30	30.00
6 Health	6.30	5.30	5.00	2.90	3.60	2.80	4.70
7 Transport	73.10	55.50	60.00	45.00	50.40	52.60	58.20
8 Communication	11.60	9.40	10.60	8.50	9.70	10.20	10.40
9 Recreation & culture	66.50	49.80	56.00	51.20	49.40	48.00	55.00
10 Education	7.80	4.60	5.70	2.90	4.20	4.10	5.40
11 Restaurants & hotels	36.80	29.40	34.70	29.10	31.10	35.00	34.10
12 Miscellaneous goods & services	38.70	31.90	32.70	25.30	27.30	31.40	31.80
1-12 All expenditure groups	397.60	323.50	348.40	291.50	313.30	335.80	342.10
13 Other expenditure items	77.40	57.00	60.00	44.20	51.80	47.00	58.20
Total expenditure	475.00	380.50	408.50	335.70	365.10	382.80	400.30
Average weekly expenditure per person (£)							
Total expenditure	205.30	170.30	172.10	146.10	156.80	142.60	168.60

Note: The commodity and service categories are not comparable with those in publications before 2001-02
1 Excluding mortgage interest payments, council tax and Northern Ireland rates

5.2 Household expenditure as a percentage of total expenditure by UK Countries and Government Office Regions 2001-02 – 2002-03

based on weighted data and including children's expenditure

	North East	North West	Yorks & the Humber	East Midlands	West Midlands	East	London
Average grossed number of households (thousands)	1,020	2,780	2,130	1,720	2,120	2,230	2,850
Total number of households in sample (over 2 years)	632	1,599	1,159	974	1,203	1,278	1,283
Total number of persons in sample (over 2 years)	1,565	3,872	2,698	2,353	3,022	3,075	3,135
Total number of adults in sample (over 2 years)	1,142	2,877	2,029	1,767	2,189	2,313	2,287
Weighted average number of persons per household	2.4	2.4	2.3	2.4	2.5	2.4	2.5
Commodity or service			**Percentage of total expenditure**				
1 **Food & non-alcoholic drinks**	*11*	*11*	*11*	*11*	*11*	*11*	*9*
2 **Alcoholic drinks, tobacco & narcotics**	*3*	*3*	*3*	*3*	*3*	*2*	*2*
3 **Clothing & footwear**	*6*	*6*	*5*	*5*	*6*	*5*	*6*
4 **Housing[1], fuel & power**	*9*	*9*	*9*	*8*	*9*	*9*	*11*
5 **Household goods & services**	*7*	*8*	*8*	*8*	*7*	*7*	*7*
6 **Health**	*1*	*1*	*1*	*1*	*1*	*2*	*1*
7 **Transport**	*15*	*14*	*15*	*15*	*15*	*16*	*13*
8 **Communication**	*3*	*2*	*3*	*2*	*3*	*3*	*3*
9 **Recreation & culture**	*15*	*15*	*15*	*15*	*14*	*14*	*12*
10 **Education**	*1*	*1*	*1*	*1*	*1*	*1*	*2*
11 **Restaurants & hotels**	*9*	*9*	*10*	*9*	*9*	*8*	*9*
12 **Miscellaneous goods & services**	*7*	*8*	*7*	*8*	*8*	*8*	*8*
1-12 **All expenditure groups**	*87*	*87*	*87*	*86*	*86*	*86*	*83*
13 **Other expenditure items**	*13*	*13*	*13*	*14*	*14*	*14*	*17*
Total expenditure	*100*	*100*	*100*	*100*	*100*	*100*	*100*

Note: The commodity and service categories are not comparable with those in publications before 2001-02
1 Excluding mortgage interest payments, council tax and Northern Ireland rates

5.2 Household expenditure as a percentage of total expenditure by UK Countries and Government Office Regions (cont.) 2001-02 – 2002-03

based on weighted data and including children's expenditure

	South East	South West	England	Wales	Scotland	Northern Ireland	United Kingdom
Average grossed number of households (thousands)	3,390	2,160	20,390	1,250	2,140	620	24,400
Total number of households in sample (over 2 years)	1,955	1,282	11,365	712	1,207	1,116	14,400
Total number of persons in sample (over 2 years)	4,746	2,925	27,391	1,644	2,759	2,914	34,708
Total number of adults in sample (over 2 years)	3,584	2,275	20,463	1,233	2,112	2,088	25,896
Weighted average number of persons per household	2.3	2.2	2.4	2.3	2.3	2.7	2.4
Commodity or service			**Percentage of total expenditure**				
1 Food & non-alcoholic drinks	9	11	10	12	12	12	11
2 Alcoholic drinks, tobacco & narcotics	2	3	3	3	4	4	3
3 Clothing & footwear	5	5	6	5	6	8	6
4 Housing[1], fuel & power	8	9	9	9	9	8	9
5 Household goods & services	7	8	7	8	7	8	7
6 Health	1	1	1	1	1	1	1
7 Transport	15	15	15	13	14	14	15
8 Communication	2	2	3	3	3	3	3
9 Recreation & culture	14	13	14	15	14	13	14
10 Education	2	1	1	1	1	1	1
11 Restaurants & hotels	8	8	8	9	9	9	9
12 Miscellaneous goods & services	8	8	8	8	7	8	8
1-12 All expenditure groups	84	85	85	87	86	88	85
13 Other expenditure items	16	15	15	13	14	12	15
Total expenditure	100	100	100	100	100	100	100

Note: The commodity and service categories are not comparable with those in publications before 2001-02
1 Excluding mortgage interest payments, council tax and Northern Ireland rates

5.3 Detailed household expenditure by UK Countries and Government Office Regions

2001-02 – 2002-03

based on weighted data and including children's expenditure

		North East	North West	Yorks & the Humber	East Midlands	West Midlands	East	London
Average grossed number of households (thousands)		1,020	2,780	2,130	1,720	2,120	2,230	2,850
Total number of households in sample (over 2 years)		632	1,599	1,159	974	1,203	1,278	1,283
Total number of persons in sample (over 2 years)		1,565	3,872	2,698	2,353	3,022	3,075	3,135
Total number of adults in sample (over 2 years)		1,142	2,877	2,029	1,767	2,189	2,313	2,287
Weighted average number of persons per household		2.4	2.4	2.3	2.4	2.5	2.4	2.5
Commodity or service		**Average weekly household expenditure (£)**						
1	**Food & non-alcoholic drinks**	**38.90**	**40.40**	**37.40**	**42.60**	**40.90**	**45.10**	**44.50**
1.1	Food	35.40	36.90	34.40	39.10	37.40	41.30	40.50
1.1.1	Bread, rice and cereals	3.60	3.80	3.50	3.90	3.80	3.80	3.90
1.1.2	Pasta products	0.30	0.30	0.30	0.30	0.20	0.40	0.40
1.1.3	Buns, cakes, biscuits etc.	2.70	2.70	2.50	2.70	2.70	3.00	2.50
1.1.4	Pastry (savoury)	0.70	0.60	0.60	0.70	0.50	0.70	0.60
1.1.5	Beef (fresh, chilled or frozen)	1.10	1.20	1.30	1.40	1.40	1.40	1.20
1.1.6	Pork (fresh, chilled or frozen)	0.50	0.60	0.60	0.50	0.60	0.60	0.50
1.1.7	Lamb (fresh, chilled or frozen)	0.40	0.60	0.40	0.50	0.60	0.60	0.80
1.1.8	Poultry (fresh, chilled or frozen)	1.30	1.40	1.30	1.50	1.50	1.70	1.70
1.1.9	Bacon and ham	0.90	1.00	0.90	0.90	0.90	0.80	0.70
1.1.10	Other meat and meat preparations	4.80	4.90	4.20	4.80	4.60	4.90	4.30
1.1.11	Fish and fish products	1.50	1.60	1.60	1.70	1.50	2.00	2.40
1.1.12	Milk	2.10	2.30	2.10	2.30	2.20	2.10	2.00
1.1.13	Cheese and curd	1.10	1.20	1.10	1.40	1.30	1.60	1.50
1.1.14	Eggs	0.40	0.40	0.30	0.40	0.40	0.40	0.50
1.1.15	Other milk products	1.30	1.20	1.10	1.50	1.20	1.50	1.40
1.1.16	Butter	0.30	0.30	0.30	0.30	0.20	0.30	0.30
1.1.17	Margarine and other vegetable fats	0.40	0.40	0.40	0.40	0.50	0.40	0.40
1.1.18	Peanut butter	0.00	0.00	0.00	0.00	[0.00]	0.00	0.00
1.1.19	Cooking oils and fats	0.10	0.20	0.20	0.20	0.20	0.20	0.30
1.1.20	Fresh fruit	1.70	2.00	1.90	2.30	2.10	2.70	2.90
1.1.21	Other fresh, chilled or frozen fruits	0.10	0.20	0.20	0.20	0.20	0.20	0.30
1.1.22	Dried fruit and nuts	0.30	0.30	0.30	0.30	0.30	0.40	0.40
1.1.23	Preserved fruit and fruit based products	0.10	0.10	0.10	0.10	0.10	0.20	0.10
1.1.24	Fresh vegetables	2.20	2.50	2.60	3.00	2.90	3.50	3.90
1.1.25	Dried vegetables and other preserved or processed vegetables	0.90	0.90	0.90	1.00	1.00	1.00	1.10
1.1.26	Potatoes	0.70	0.80	0.70	0.80	0.80	0.80	0.80
1.1.27	Other tubers and products of tuber vegetables	1.30	1.20	1.00	1.20	1.30	1.10	0.90
1.1.28	Sugar and sugar products	0.20	0.20	0.20	0.30	0.30	0.30	0.20
1.1.28	Jams, marmalades	0.20	0.20	0.20	0.20	0.20	0.20	0.20
1.1.30	Chocolate	1.40	1.30	1.20	1.40	1.30	1.30	1.10
1.1.31	Confectionery products	0.70	0.60	0.60	0.60	0.60	0.60	0.50
1.1.32	Edible ices and ice cream	0.50	0.40	0.40	0.50	0.50	0.60	0.50
1.1.33	Other food products	1.60	1.70	1.70	1.90	1.60	1.90	1.90
1.2	Non-alcoholic drinks	3.50	3.50	3.00	3.50	3.50	3.80	4.00
1.2.1	Coffee	0.50	0.50	0.50	0.60	0.50	0.50	0.40
1.2.2	Tea	0.40	0.50	0.40	0.50	0.50	0.50	0.40
1.2.3	Cocoa and powdered chocolate	0.10	0.10	0.10	0.10	0.10	0.10	0.10
1.2.4	Fruit and vegetable juices, mineral waters	0.90	1.00	0.80	1.10	0.90	1.30	1.60
1.2.5	Soft drinks	1.60	1.40	1.20	1.30	1.40	1.40	1.50

Note: The commodity and service categories are not comparable with those in publications before 2001-02
The numbering system is sequential, it does not use actual COICOP codes

5.3 Detailed household expenditure by UK Countries and Government Office Regions (cont.)

2001-02 – 2002-03

based on weighted data and including children's expenditure

			South East	South West	England	Wales	Scotland	Northern Ireland	United Kingdom
Average grossed number of households (thousands)			3,390	2,160	20,390	1,250	2,140	620	24,400
Total number of households in sample (over 2 years)			1,955	1,282	11,365	712	1,207	1,116	14,400
Total number of persons in sample (over 2 years)			4,746	2,925	27,391	1,644	2,759	2,914	34,708
Total number of adults in sample (over 2 years)			3,584	2,275	20,463	1,233	2,112	2,088	25,896
Weighted average number of persons per household			2.3	2.2	2.4	2.3	2.3	2.7	2.4
Commodity or service			**Average weekly household expenditure (£)**						
1	**Food & non-alcoholic drinks**		**44.90**	**42.20**	**42.30**	**39.60**	**42.20**	**46.20**	**42.20**
1.1	Food		41.10	38.80	38.70	36.40	38.20	42.00	38.60
	1.1.1	Bread, rice and cereals	3.80	3.60	3.70	3.50	4.00	4.80	3.80
	1.1.2	Pasta products	0.30	0.30	0.30	0.30	0.40	0.30	0.30
	1.1.3	Buns, cakes, biscuits etc.	2.80	2.80	2.70	2.50	2.70	3.40	2.70
	1.1.4	Pastry (savoury)	0.70	0.60	0.60	0.50	0.60	0.60	0.60
	1.1.5	Beef (fresh, chilled or frozen)	1.30	1.30	1.30	1.10	1.60	2.60	1.30
	1.1.6	Pork (fresh, chilled or frozen)	0.60	0.60	0.60	0.50	0.50	0.60	0.60
	1.1.7	Lamb (fresh, chilled or frozen)	0.60	0.60	0.60	0.60	0.30	0.30	0.60
	1.1.8	Poultry (fresh, chilled or frozen)	1.60	1.40	1.50	1.30	1.50	1.80	1.50
	1.1.9	Bacon and ham	0.80	0.90	0.80	0.90	0.90	1.20	0.90
	1.1.10	Other meat and meat preparations	4.90	4.20	4.60	4.90	5.20	5.10	4.70
	1.1.11	Fish and fish products	2.00	1.80	1.80	1.70	1.60	1.40	1.80
	1.1.12	Milk	2.00	2.20	2.10	2.10	2.10	2.90	2.10
	1.1.13	Cheese and curd	1.60	1.50	1.40	1.10	1.30	1.10	1.40
	1.1.14	Eggs	0.40	0.40	0.40	0.40	0.40	0.40	0.40
	1.1.15	Other milk products	1.60	1.50	1.40	1.20	1.30	1.30	1.40
	1.1.16	Butter	0.30	0.30	0.30	0.30	0.30	0.40	0.30
	1.1.17	Margarine and other vegetable fats	0.40	0.40	0.40	0.40	0.40	0.50	0.40
	1.1.18	Peanut butter	0.00	0.00	0.00	[0.00]	0.00	0.00	0.00
	1.1.19	Cooking oils and fats	0.20	0.20	0.20	0.20	0.10	0.20	0.20
	1.1.20	Fresh fruit	2.70	2.70	2.40	2.00	2.00	1.90	2.30
	1.1.21	Other fresh, chilled or frozen fruits	0.30	0.20	0.20	0.20	0.20	0.20	0.20
	1.1.22	Dried fruit and nuts	0.40	0.40	0.40	0.30	0.20	0.20	0.30
	1.1.23	Preserved fruit and fruit based products	0.20	0.20	0.10	0.20	0.10	0.20	0.10
	1.1.24	Fresh vegetables	3.60	3.10	3.10	2.60	2.30	2.20	3.00
	1.1.25	Dried vegetables and other preserved or processed vegetables	1.10	1.00	1.00	0.90	1.00	1.00	1.00
	1.1.26	Potatoes	0.80	0.70	0.80	0.90	0.70	1.20	0.80
	1.1.27	Other tubers and products of tuber vegetables	1.10	1.10	1.10	1.30	1.40	1.40	1.20
	1.1.28	Sugar and sugar products	0.30	0.30	0.30	0.30	0.20	0.20	0.30
	1.1.28	Jams, marmalades	0.20	0.20	0.20	0.20	0.20	0.30	0.20
	1.1.30	Chocolate	1.40	1.30	1.30	1.20	1.30	1.40	1.30
	1.1.31	Confectionery products	0.50	0.50	0.60	0.60	0.60	0.80	0.60
	1.1.32	Edible ices and ice cream	0.60	0.50	0.50	0.50	0.50	0.50	0.50
	1.1.33	Other food products	2.00	1.80	1.80	1.60	1.80	1.90	1.80
1.2	Non-alcoholic drinks		3.90	3.40	3.60	3.20	4.00	4.20	3.60
	1.2.1	Coffee	0.50	0.60	0.50	0.50	0.50	0.40	0.50
	1.2.2	Tea	0.50	0.50	0.50	0.50	0.40	0.50	0.50
	1.2.3	Cocoa and powdered chocolate	0.10	0.10	0.10	0.10	0.00	0.00	0.10
	1.2.4	Fruit and vegetable juices, mineral waters	1.30	1.10	1.20	0.80	1.10	1.10	1.10
	1.2.5	Soft drinks	1.40	1.00	1.30	1.30	2.00	2.10	1.40

Note: The commodity and service categories are not comparable with those in publications before 2001-02
The numbering system is sequential, it does not use actual COICOP codes

5.3 Detailed household expenditure by UK Countries and Government Office Regions (cont.) 2001-02 – 2002-03

based on weighted data and including children's expenditure

Commodity or service			North East	North West	Yorks & the Humber	East Midlands	West Midlands	East	London
			Average weekly household expenditure (£)						
2	**Alcoholic drink, tobacco & narcotics**		**10.90**	**12.10**	**11.00**	**11.70**	**11.60**	**9.90**	**10.30**
2.1	Alcoholic drinks		5.60	6.40	5.30	6.00	5.60	5.60	5.70
	2.1.1	Spirits and liqueurs (brought home)	1.00	1.40	0.90	1.10	1.30	1.00	1.00
	2.1.2	Wines, fortified wines (brought home)	2.20	2.80	2.30	3.00	2.50	2.80	3.30
	2.1.3	Beer, lager, ciders and Perry (brought home)	2.30	2.00	2.00	1.80	1.70	1.70	1.30
	2.1.4	Alcopops (brought home)	0.10	0.20	0.10	0.10	0.10	0.20	0.10
2.2	Tobacco and narcotics		5.30	5.70	5.70	5.70	5.90	4.30	4.60
	2.2.1	Cigarettes	4.80	5.00	5.20	5.10	5.30	3.90	4.00
	2.2.2	Cigars, other tobacco products and narcotics	0.50	0.70	0.50	0.60	0.70	0.40	0.60
3	**Clothing & footwear**		**20.50**	**23.20**	**18.80**	**19.40**	**22.00**	**22.60**	**27.50**
3.1	Clothing		16.70	18.60	15.30	15.80	18.10	19.00	22.40
	3.1.1	Men's outer garments	4.30	4.60	4.20	3.50	4.50	4.80	5.70
	3.1.2	Men's under garments	0.30	0.40	0.30	0.50	0.40	0.50	0.50
	3.1.3	Women's outer garments	6.50	7.80	6.20	6.90	7.70	8.00	9.50
	3.1.4	Women's under garments	1.10	1.30	1.10	1.20	1.50	1.50	1.20
	3.1.5	Boys' outer garments (5-15)	1.20	0.80	0.90	0.70	0.90	0.70	0.90
	3.1.6	Girls' outer garments (5-15)	1.30	1.20	0.80	1.20	1.20	1.20	1.40
	3.1.7	Infants' outer garments (under 5)	0.70	0.80	0.60	0.50	0.50	0.60	0.80
	3.1.8	Children's under garments (under 16)	0.40	0.50	0.40	0.30	0.40	0.40	0.40
	3.1.9	Accessories	0.50	0.60	0.50	0.70	0.60	0.70	0.90
	3.1.10	Haberdashery, clothing materials and clothing hire	0.20	0.20	0.20	0.10	0.20	0.20	0.50
	3.1.11	Dry cleaners, laundry and dyeing	[0.10]	0.30	0.20	0.20	0.20	0.30	0.50
3.2	Footwear		3.80	4.60	3.50	3.60	3.90	3.60	5.10
4	**Housing,[1] fuel & power**		**29.30**	**32.10**	**31.40**	**31.40**	**31.20**	**37.40**	**55.50**
4.1	Actual rentals for housing		19.30	19.80	18.40	13.20	18.20	20.10	45.60
	4.1.1	Gross rent	19.30	19.50	18.40	13.10	18.10	20.00	45.60
	4.1.2	*less* housing benefit, rebates & allowances rec'd	11.40	11.40	8.80	6.20	9.50	6.80	14.10
	4.1.3	Net rent	7.90	8.10	9.60	7.00	8.60	13.30	31.50
	4.1.4	Second dwelling rent	..	..	..	..	..	..	..
4.2	Maintenance and repair of dwelling		5.20	6.80	6.00	7.70	6.20	6.80	7.50
4.3	Water supply and miscellaneous services relating to the dwelling		4.40	5.30	4.80	4.70	4.90	5.60	5.90
4.4	Electricity, gas and other fuels		11.70	11.60	11.00	12.00	11.40	11.70	10.50
	4.4.1	Electricity	5.60	5.50	5.30	5.50	5.50	5.70	5.10
	4.4.2	Gas	5.40	5.80	5.10	5.80	5.40	4.70	5.30
	4.4.3	Other fuels	0.70	0.30	0.60	0.70	0.50	1.40	0.10

Note: The commodity and service categories are not comparable with those in previous publications
The numbering system is sequential, it does not use actual COICOP codes
1 Excluding mortgage interest payments, council tax and Northern Ireland rates

5.3 Detailed household expenditure by UK Countries and Government Office Regions (cont.)

2001-02 – 2002-03

based on weighted data and including children's expenditure

	Commodity or service	South East	South West	England	Wales	Scotland	Northern Ireland	United Kingdom
		Average weekly household expenditure (£)						
2	**Alcoholic drink, tobacco & narcotics**	**11.10**	**10.80**	**11.00**	**11.20**	**13.40**	**13.50**	**11.30**
2.1	Alcoholic drinks	6.00	5.90	5.80	5.80	6.30	5.10	5.90
2.1.1	Spirits and liqueurs (brought home)	1.10	1.30	1.10	1.20	2.00	1.30	1.20
2.1.2	Wines, fortified wines (brought home)	3.30	3.10	2.90	2.40	2.50	2.10	2.80
2.1.3	Beer, lager, ciders and Perry (brought home)	1.50	1.40	1.70	2.00	1.60	1.50	1.70
2.1.4	Alcopops (brought home)	0.20	0.10	0.10	0.10	0.20	0.20	0.10
2.2	Tobacco and narcotics	5.00	4.90	5.20	5.40	7.10	8.50	5.40
2.2.1	Cigarettes	4.30	4.20	4.60	4.70	6.40	8.00	4.80
2.2.2	Cigars, other tobacco products and narcotics	0.70	0.70	0.60	0.70	0.70	0.50	0.60
3	**Clothing & footwear**	**25.40**	**18.00**	**22.50**	**17.80**	**23.10**	**30.20**	**22.50**
3.1	Clothing	21.00	15.00	18.50	14.20	18.90	23.80	18.40
3.1.1	Men's outer garments	5.00	3.80	4.60	3.00	4.50	6.30	4.50
3.1.2	Men's under garments	0.50	0.40	0.40	0.20	0.40	0.50	0.40
3.1.3	Women's outer garments	9.90	6.70	8.00	6.90	8.00	9.60	8.00
3.1.4	Women's under garments	1.40	1.20	1.30	0.90	1.30	1.40	1.30
3.1.5	Boys' outer garments (5-15)	0.80	0.50	0.80	0.60	1.00	1.60	0.80
3.1.6	Girls' outer garments (5-15)	1.10	0.90	1.10	0.90	1.40	1.70	1.20
3.1.7	Infants' outer garments (under 5)	0.50	0.30	0.60	0.60	0.60	0.90	0.60
3.1.8	Children's under garments (under 16)	0.40	0.30	0.40	0.30	0.40	0.50	0.40
3.1.9	Accessories	0.80	0.50	0.70	0.40	0.70	0.70	0.70
3.1.10	Haberdashery, clothing materials and clothing	0.30	0.20	0.30	0.10	0.20	0.20	0.20
3.1.11	Dry cleaners, laundry and dyeing	0.40	0.30	0.30	0.10	0.30	0.30	0.30
3.2	Footwear	4.30	3.00	4.00	3.60	4.20	6.40	4.10
4	**Housing[1], fuel & power**	**40.20**	**36.10**	**37.30**	**31.90**	**32.40**	**30.30**	**36.50**
4.1	Actual rentals for housing	20.50	16.90	22.30	16.40	19.70	14.90	21.60
4.1.1	Gross rent	20.40	16.90	22.30	16.40	19.60	14.90	21.50
4.1.2	*less* housing benefit, rebates & allowances rec	6.70	6.50	9.10	8.50	10.10	8.70	9.10
4.1.3	Net rent	13.70	10.40	13.20	7.90	9.60	6.10	12.40
4.1.4	Second dwelling rent	..	..	0.10	..	..	..	0.10
4.2	Maintenance and repair of dwelling	8.90	7.80	7.20	6.30	5.30	6.60	7.00
4.3	Water supply and miscellaneous services relating to the dwelling	5.90	5.90	5.40	5.70	5.10	0.30	5.30
4.4	Electricity, gas and other fuels	11.70	12.10	11.50	11.90	12.30	17.30	11.70
4.4.1	Electricity	5.90	6.30	5.60	5.80	7.00	7.90	5.80
4.4.2	Gas	5.20	4.60	5.20	5.40	4.70	0.60	5.10
4.4.3	Other fuels	0.60	1.10	0.60	0.80	0.70	8.80	0.90

Note: The commodity and service categories are not comparable with those in publications before 2001-02
The numbering system is sequential, it does not use actual COICOP codes
1 Excluding mortgage interest payments, council tax and Northern Ireland rates

5.3 Detailed household expenditure by UK Countries and Government Office Regions (cont.)

2001-02 – 2002-03

based on weighted data and including children's expenditure

		North East	North West	Yorks & the Humber	East Midlands	West Midlands	East	London
Commodity or service		Average weekly household expenditure (£)						
5	**Household goods & services**	**23.60**	**29.20**	**26.90**	**30.40**	**27.10**	**30.40**	**33.80**
5.1	Furniture and furnishings, carpets and other floor coverings	11.70	16.20	13.60	15.10	14.60	15.50	18.40
	5.1.1 Furniture and furnishings	8.30	12.40	10.10	10.70	10.80	12.20	14.90
	5.1.2 Floor coverings	3.40	3.80	3.40	4.20	3.80	3.30	3.50
	5.1.3 Repair of furniture, furnishings and floor coverings	..	..	..	..	..	..	..
5.2	Household textiles	1.90	1.90	1.80	2.10	1.90	2.20	2.10
5.3	Household appliances	3.10	3.00	3.40	4.70	2.90	3.30	3.90
5.4	Glassware, tableware and household utensils	1.30	1.30	1.60	1.40	1.20	1.70	1.50
5.5	Tools and equipment for house and garden	1.90	2.20	2.20	2.60	2.30	2.80	2.80
5.6	Goods and services for routine household maintenance	3.70	4.50	4.30	4.40	4.30	4.90	5.20
	5.6.1 Cleaning materials	1.90	2.00	1.90	2.00	2.00	2.30	1.90
	5.6.2 Household goods and hardware	1.00	0.90	0.90	1.20	1.00	1.20	1.10
	5.6.3 Domestic services, carpet cleaning	0.90	1.60	1.50	1.20	1.30	1.40	2.10
6	**Health**	**3.20**	**3.40**	**3.80**	**3.90**	**3.50**	**6.80**	**6.40**
6.1	Medical products, appliances and equipment	2.40	2.40	2.90	2.60	2.20	3.70	2.70
	6.1.1 Medicines, prescriptions and healthcare products	1.10	1.40	1.60	1.40	1.30	1.80	1.60
	6.1.2 Spectacles, lenses, accessories and repairs	1.30	1.00	1.30	1.10	0.90	1.80	1.00
	6.1.3 Non-optical appliances and equipment (e.g. wheelchairs, batteries for hearing aids, etc.)	..	..	..	..	..	..	..
6.2	Hospital services	0.80	0.90	0.90	1.30	1.30	3.10	3.70
7	**Transport**	**49.60**	**53.70**	**52.20**	**59.10**	**53.10**	**69.40**	**62.20**
7.1	Purchase of vehicles	22.30	25.10	26.50	28.80	23.70	33.10	23.40
	7.1.1 Purchase of new cars and vans	9.50	10.00	11.80	10.10	8.90	15.00	10.80
	7.1.2 Purchase of second hand cars or vans	12.30	14.50	14.30	17.70	14.40	17.30	11.60
	7.1.3 Purchase of motorcycles and other vehicles	[0.50]	0.60	[0.30]	1.00	[0.50]	0.80	1.00
7.2	Operation of personal transport	19.50	21.20	19.70	25.00	23.50	27.90	20.90
	7.2.1 Spares and accessories	1.50	1.60	1.30	1.90	1.90	2.50	1.90
	7.2.2 Petrol, diesel and other motor oils	13.20	13.80	12.30	16.00	15.50	17.20	11.00
	7.2.3 Repairs and servicing	3.50	4.10	4.40	5.40	4.50	5.70	6.00
	7.2.4 Other motoring costs	1.30	1.70	1.70	1.70	1.60	2.40	2.00
7.3	Transport services	7.90	7.40	6.10	5.40	5.90	8.50	17.90
	7.3.1 Rail and tube fares	1.10	1.00	0.80	0.90	0.90	3.60	4.00
	7.3.2 Bus and coach fares	1.60	1.60	1.80	1.20	1.30	0.90	2.20
	7.3.3 Combined fares	[0.30]	[0.10]	[0.20]	..	..	0.60	6.10
	7.3.4 Other travel and transport	4.90	4.70	3.30	3.20	3.70	3.30	5.60

Note: The commodity and service categories are not comparable with those in publications before 2001-02
The numbering system is sequential, it does not use actual COICOP codes

5.3 Detailed household expenditure by UK Countries and Government Office Regions (cont.)

2001-02 – 2002-03

based on weighted data and including children's expenditure

		South East	South West	England	Wales	Scotland	Northern Ireland	United Kingdom
Commodity or service		**Average weekly household expenditure (£)**						
5	**Household goods & services**	**35.40**	**30.60**	**30.50**	**26.20**	**26.60**	**31.30**	**30.00**
5.1	Furniture and furnishings, carpets and other floor coverings	17.80	15.60	15.90	12.70	12.30	16.80	15.40
5.1.1	Furniture and furnishings	13.30	11.70	12.10	9.90	9.80	12.70	11.80
5.1.2	Floor coverings	4.50	4.00	3.80	2.90	2.50	4.00	3.70
5.1.3	Repair of furniture, furnishings and floor coveri	..	..	[0.00]	..	..	..	[0.00]
5.2	Household textiles	2.40	1.90	2.00	2.00	2.40	1.90	2.10
5.3	Household appliances	4.20	3.50	3.60	4.00	4.50	3.10	3.70
5.4	Glassware, tableware and household utensils	2.40	2.20	1.70	1.30	1.30	1.50	1.60
5.5	Tools and equipment for house and garden	3.30	2.60	2.60	2.50	2.30	3.90	2.60
5.6	Goods and services for routine household maintenance	5.20	4.80	4.70	3.70	3.70	4.10	4.60
5.6.1	Cleaning materials	2.20	1.90	2.00	2.00	1.80	2.10	2.00
5.6.2	Household goods and hardware	1.20	1.20	1.10	0.90	1.00	1.00	1.10
5.6.3	Domestic services, carpet cleaning	1.80	1.70	1.60	0.80	0.90	1.00	1.50
6	**Health**	**6.30**	**5.30**	**5.00**	**2.90**	**3.60**	**2.80**	**4.70**
6.1	Medical products, appliances and equipment	3.80	3.30	3.00	1.90	2.50	2.00	2.90
6.1.1	Medicines, prescriptions and healthcare produ	2.00	1.40	1.60	1.20	1.10	1.20	1.50
6.1.2	Spectacles, lenses, accessories and repairs	1.80	1.60	1.30	0.60	1.40	0.80	1.30
6.1.3	Non-optical appliances and equipment (e.g. wheelchairs, batteries for hearing aids, etc.)	..	..	0.10	..	..	..	0.10
6.2	Hospital services	2.40	2.00	2.00	1.00	1.10	0.90	1.80
7	**Transport**	**73.10**	**55.50**	**60.00**	**45.00**	**50.40**	**52.60**	**58.20**
7.1	Purchase of vehicles	33.60	24.20	27.20	18.10	21.50	21.60	26.10
7.1.1	Purchase of new cars and vans	15.20	10.00	11.60	5.80	8.10	11.50	11.00
7.1.2	Purchase of second hand cars or vans	17.10	13.10	14.80	11.80	12.90	9.50	14.30
7.1.3	Purchase of motorcycles and other vehicles	1.20	1.10	0.80	..	[0.40]	[0.60]	0.80
7.2	Operation of personal transport	29.90	26.20	24.20	21.30	20.70	23.40	23.70
7.2.1	Spares and accessories	3.00	2.40	2.10	1.50	1.80	1.60	2.00
7.2.2	Petrol, diesel and other motor oils	17.70	15.90	14.80	14.10	13.50	16.90	14.70
7.2.3	Repairs and servicing	6.90	5.60	5.30	4.20	3.90	3.60	5.10
7.2.4	Other motoring costs	2.30	2.20	1.90	1.50	1.50	1.20	1.90
7.3	Transport services	9.50	5.10	8.60	5.50	8.30	7.60	8.40
7.3.1	Rail and tube fares	2.90	1.10	2.00	0.70	1.30	0.30	1.80
7.3.2	Bus and coach fares	1.00	1.00	1.40	1.20	2.10	1.10	1.40
7.3.3	Combined fares	0.50	..	1.10	..	[0.20]	..	0.90
7.3.4	Other travel and transport	5.10	3.00	4.20	3.70	4.70	6.20	4.30

Note: The commodity and service categories are not comparable with those in publications before 2001-02
The numbering system is sequential, it does not use actual COICOP codes

5.3 Detailed household expenditure by UK Countries and Government Office Regions (cont.)

2001-02 – 2002-03

based on weighted data and including children's expenditure

	Commodity or service	North East	North West	Yorks & the Humber	East Midlands	West Midlands	East	London
		Average weekly household expenditure (£)						
8	**Communication**	**9.80**	**9.30**	**9.00**	**9.60**	**9.70**	**10.90**	**14.50**
8.1	Postal services	0.50	0.40	0.40	0.40	0.40	0.50	0.60
8.2	Telephone and telefax equipment	1.20	0.50	0.50	0.70	0.70	0.70	1.00
8.3	Telephone and telefax services	8.20	8.30	8.10	8.50	8.70	9.80	12.90
9	**Recreation & culture**	**50.20**	**55.40**	**51.40**	**56.50**	**51.60**	**57.30**	**56.70**
9.1	Audio-visual, photographic and information processing equipment	6.50	7.50	7.50	7.60	8.60	7.50	9.40
9.1.1	Audio equipment and accessories, CD players	2.20	2.60	2.70	2.30	2.00	2.80	2.80
9.1.2	TV, video and computers	4.10	4.20	4.20	4.30	6.00	3.90	5.70
9.1.3	Photographic, cinematographic & optical equip't	0.20	0.70	0.60	1.10	0.60	0.80	0.90
9.2	Other major durables for recreation and culture	[0.50]	..	..	..	1.00	1.60	[0.70]
9.3	Other recreational items and equipment, gardens and pets	8.90	9.00	7.90	10.20	8.90	11.00	8.50
9.3.1	Games, toys and hobbies	2.50	2.10	1.60	2.30	2.00	2.50	2.40
9.3.2	Computer software and games	1.00	1.10	1.20	0.70	1.20	1.00	0.70
9.3.3	Equipment for sport, camping and open-air recreation	1.70	0.70	0.50	1.40	0.80	0.70	0.60
9.3.4	Horticultural goods, garden equipment and plants	2.00	2.30	2.20	3.10	2.30	3.30	2.80
9.3.5	Pets and pet food	1.70	2.80	2.50	2.70	2.50	3.60	2.00
9.4	Recreational and cultural services	18.10	16.90	14.90	16.60	15.10	17.30	19.30
9.4.1	Sports admissions, subscriptions and leisure class fees	4.20	4.70	3.80	5.00	4.10	5.60	7.90
9.4.2	Cinema, theatre and museums etc.	1.40	1.40	1.10	1.60	1.50	1.50	2.40
9.4.3	TV, video, satellite rental, cable subscriptions, TV licences and the Internet	5.00	5.20	4.20	5.00	4.40	4.90	4.80
9.4.4	Miscellaneous entertainments	1.10	1.00	0.80	0.70	0.80	1.30	1.00
9.4.5	Development of film, deposit for film development, passport photos, holiday and school photos	0.40	0.60	0.30	0.30	0.50	0.60	0.40
9.4.6	Gambling payments	6.10	4.00	4.70	3.90	3.80	3.50	2.90
9.5	Newspapers, books and stationery	5.90	6.20	5.30	6.50	5.50	6.60	7.30
9.5.1	Books, diaries, address books, cards etc.	3.00	3.20	2.60	3.40	2.80	3.70	4.40
9.5.2	Newspapers	2.00	1.90	1.70	2.00	1.70	1.80	1.80
9.5.3	Magazines and periodicals	1.00	1.10	1.00	1.10	0.90	1.20	1.20
9.6	Package holidays	10.30	13.40	13.10	11.30	12.50	13.30	11.50
9.6.1	Package holidays - UK	[0.40]	0.80	1.30	1.10	0.80	1.00	1.00
9.6.2	Package holidays - abroad	9.90	12.60	11.80	10.20	11.70	12.30	10.50
10	**Education**	**3.40**	**5.20**	**3.70**	**3.80**	**2.80**	**5.30**	**10.30**
10.1	Education fees	3.30	4.80	3.40	3.50	2.70	5.00	9.80
10.2	Payments for school trips, other ad-hoc expenditure	[0.10]	0.30	0.30	0.30	0.10	0.30	0.50

Note: The commodity and service categories are not comparable with those in publications before 2001-02
The numbering system is sequential, it does not use actual COICOP codes

5.3 Detailed household expenditure by UK Countries and Government Office Regions (cont.) 2001-02 – 2002-03

based on weighted data and including children's expenditure

Commodity or service		South East	South West	England	Wales	Scotland	Northern Ireland	United Kingdom
		Average weekly household expenditure (£)						
8	**Communication**	**11.60**	**9.40**	**10.60**	**8.50**	**9.70**	**10.20**	**10.40**
8.1	Postal services	0.60	0.60	0.50	0.30	0.40	0.30	0.50
8.2	Telephone and telefax equipment	0.60	0.40	0.70	0.30	0.60	0.40	0.60
8.3	Telephone and telefax services	10.40	8.40	9.50	7.90	8.70	9.50	9.30
9	**Recreation & culture**	**66.50**	**49.80**	**56.00**	**51.20**	**49.40**	**48.00**	**55.00**
9.1	Audio-visual, photographic and information processing equipment	10.10	5.80	8.10	10.10	6.50	5.60	8.00
9.1.1	Audio equipment and accessories, CD players	3.50	2.10	2.70	1.80	2.30	1.50	2.50
9.1.2	TV, video and computers	5.50	3.10	4.70	7.60	3.60	3.80	4.70
9.1.3	Photographic, cinematographic and optical equ	1.00	0.70	0.80	0.70	0.60	0.20	0.70
9.2	Other major durables for recreation and culture	2.30	[0.50]	1.80	[0.80]	0.90	2.40	1.70
9.3	Other recreational items and equipment, gardens and pets	13.20	10.70	10.00	9.40	7.70	7.90	9.70
9.3.1	Games, toys and hobbies	2.40	2.20	2.20	2.10	1.90	2.10	2.20
9.3.2	Computer software and games	1.20	0.80	1.00	1.40	1.20	1.00	1.00
9.3.3	Equipment for sport, camping and open-air recreation	1.40	1.00	0.90	1.00	0.90	0.70	0.90
9.3.4	Horticultural goods, garden equipment and pla	4.30	3.20	2.90	2.20	1.90	2.30	2.80
9.3.5	Pets and pet food	3.90	3.60	2.90	2.70	1.90	1.80	2.80
9.4	Recreational and cultural services	17.70	15.20	16.90	14.30	16.80	15.50	16.70
9.4.1	Sports admissions, subscriptions and leisure class fees	6.10	5.00	5.40	3.50	5.00	4.80	5.20
9.4.2	Cinema, theatre and museums etc.	2.00	1.50	1.70	1.00	1.90	1.50	1.60
9.4.3	TV, video, satellite rental, cable subscriptions, TV licences and the Internet	4.90	4.30	4.70	5.10	4.80	4.10	4.80
9.4.4	Miscellaneous entertainments	1.20	0.90	1.00	0.70	0.80	1.20	1.00
9.4.5	Development of film, deposit for film development, passport photos, holiday and school photos	0.50	0.40	0.40	0.30	0.30	0.30	0.40
9.4.6	Gambling payments	3.00	3.10	3.70	3.70	4.00	3.50	3.70
9.5	Newspapers, books and stationery	7.00	6.70	6.40	5.70	6.50	5.90	6.40
9.5.1	Books, diaries, address books, cards etc.	4.10	3.60	3.50	2.90	3.20	2.60	3.40
9.5.2	Newspapers	1.80	1.90	1.80	1.80	2.40	2.40	1.90
9.5.3	Magazines and periodicals	1.10	1.20	1.10	0.90	0.90	1.00	1.10
9.6	Package holidays	16.30	10.80	12.90	11.00	11.00	10.60	12.60
9.6.1	Package holidays - UK	0.70	0.70	0.90	[0.80]	[0.30]	0.70	0.80
9.6.2	Package holidays - abroad	15.50	10.00	12.00	10.20	10.70	9.90	11.70
10	**Education**	**7.80**	**4.60**	**5.70**	**2.90**	**4.20**	**4.10**	**5.40**
10.1	Education fees	7.50	4.40	5.40	2.80	4.10	3.70	5.10
10.2	Payments for school trips, other ad-hoc expenditure	0.30	0.20	0.30	[0.10]	[0.10]	0.40	0.30

Note: The commodity and service categories are not comparable with those in publications before 2001-02
The numbering system is sequential, it does not use actual COICOP codes

5.3 Detailed household expenditure by UK Countries and Government Office Regions (cont.) 2001-02 – 2002-03

based on weighted data and including children's expenditure

Commodity or service		North East	North West	Yorks & the Humber	East Midlands	West Midlands	East	London
		Average weekly household expenditure (£)						
11	**Restaurants & hotels**	**32.00**	**32.30**	**33.80**	**34.50**	**31.40**	**33.60**	**43.80**
11.1	Catering services	29.30	28.80	30.10	29.70	27.30	28.30	37.80
11.1.1	Restaurant and café meals	9.00	9.80	9.40	11.10	9.20	12.00	15.20
11.1.2	Alcoholic drinks (away from home)	10.50	9.40	9.20	9.30	8.50	7.10	9.60
11.1.3	Take away meals eaten at home	3.80	3.30	3.60	3.40	3.80	3.30	4.20
11.1.4	Other take-away and snack food	4.00	4.20	3.80	3.60	3.50	3.80	6.30
11.1.5	Contract catering (food)	..	[0.40]	..	..	..	..	..
11.1.6	Canteens	1.90	1.70	1.60	2.00	1.80	1.90	2.20
11.2	Accommodation services	2.70	3.50	3.80	4.80	4.20	5.30	6.00
11.2.1	Holiday in the UK	1.80	2.20	2.20	2.60	2.80	2.60	2.10
11.2.2	Holiday abroad	0.90	1.40	1.50	2.10	1.30	2.60	3.90
11.2.3	Room hire	..	..	..	..	..	..	..
12	**Miscellaneous goods & services**	**24.60**	**31.50**	**25.50**	**30.20**	**28.10**	**34.30**	**39.40**
12.1	Personal care	7.50	8.60	7.30	9.00	7.60	9.10	9.50
12.1.1	Hairdressing, beauty treatment	1.70	2.80	2.30	2.50	2.20	2.70	2.80
12.1.2	Toilet paper	0.60	0.60	0.60	0.70	0.70	0.70	0.70
12.1.3	Toiletries and soap	1.70	1.70	1.60	2.00	1.70	2.00	2.30
12.1.4	Baby toiletries and accessories (disposable)	0.60	0.50	0.50	0.60	0.50	0.60	0.70
12.1.5	Hair products, cosmetics and electrical personal appliances	2.80	3.00	2.40	3.10	2.60	3.10	3.10
12.2	Personal effects	2.20	3.10	2.20	1.90	2.30	2.80	4.10
12.3	Social protection	1.90	2.40	1.80	1.90	1.60	4.20	4.60
12.4	Insurance	10.40	13.50	11.00	13.70	13.00	13.80	15.80
12.4.1	Household insurances - structural, contents and appliances	3.90	4.40	3.70	4.70	4.00	4.40	5.30
12.4.2	Medical insurance premiums	0.40	1.00	0.70	1.00	1.10	1.50	2.20
12.4.3	Vehicle insurance including boat insurance	6.10	8.00	6.40	7.80	7.70	7.80	8.10
12.4.4	Non-package holiday, other travel insurance	..	[0.10]	..	[0.20]	..	[0.10]	[0.10]
12.5	Other services n.e.c	2.60	3.80	3.20	3.80	3.60	4.40	5.40
12.5.1	Moving house	1.10	1.60	1.50	2.30	1.60	3.00	3.40
12.5.2	Bank, building society, post office, credit card charges	0.20	0.30	0.30	0.30	0.30	0.40	0.50
12.5.3	Other services and professional fees	1.40	1.90	1.40	1.20	1.60	1.00	1.60
1-12	**All expenditure groups**	**296.10**	**327.60**	**305.00**	**333.00**	**312.90**	**362.90**	**404.80**

Note: The commodity and service categories are not comparable with those in publications before 2001-02
The numbering system is sequential, it does not use actual COICOP codes

5.3 Detailed household expenditure by UK Countries and Government Office Regions (cont.) 2001-02 – 2002-03

based on weighted data and including children's expenditure

	Commodity or service	South East	South West	England	Wales	Scotland	Northern Ireland	United Kingdom
		Average weekly household expenditure (£)						
11	**Restaurants & hotels**	**36.80**	**29.40**	**34.70**	**29.10**	**31.10**	**35.00**	**34.10**
11.1	Catering services	30.70	24.80	30.00	25.80	27.70	33.10	29.60
11.1.1	Restaurant and café meals	12.50	10.70	11.30	8.60	9.40	11.40	11.00
11.1.2	Alcoholic drinks (away from home)	8.80	7.30	8.80	8.80	7.90	8.80	8.70
11.1.3	Take away meals eaten at home	3.50	2.40	3.50	3.20	3.70	5.30	3.50
11.1.4	Other take-away and snack food	4.00	2.80	4.10	3.50	4.50	5.00	4.10
11.1.5	Contract catering (food)	..	..	0.50	..	..	..	0.40
11.1.6	Canteens	1.70	1.40	1.80	1.60	1.90	2.50	1.80
11.2	Accommodation services	6.10	4.60	4.80	3.40	3.40	1.90	4.50
11.2.1	Holiday in the UK	2.70	2.30	2.40	1.90	1.70	0.60	2.30
11.2.2	Holiday abroad	3.10	2.20	2.30	1.50	1.60	1.30	2.20
11.2.3	Room hire	..	..	0.10	..	..	..	0.10
12	**Miscellaneous goods & services**	**38.70**	**31.90**	**32.70**	**25.30**	**27.30**	**31.40**	**31.80**
12.1	Personal care	9.90	8.50	8.70	7.20	8.20	9.50	8.60
12.1.1	Hairdressing, beauty treatment	3.40	2.90	2.70	1.70	2.20	2.90	2.60
12.1.2	Toilet paper	0.70	0.70	0.70	0.70	0.70	0.80	0.70
12.1.3	Toiletries and soap	2.10	1.90	1.90	1.70	1.90	2.00	1.90
12.1.4	Baby toiletries and accessories (disposable)	0.50	0.50	0.50	0.40	0.50	0.80	0.50
12.1.5	Hair products, cosmetics and electrical personal appliances	3.20	2.60	2.90	2.60	2.80	3.10	2.90
12.3	Personal effects	4.10	2.40	3.00	2.30	2.60	2.10	2.90
12.3	Social protection	2.10	2.20	2.60	2.40	2.40	2.60	2.60
12.4	Insurance	16.80	14.20	14.00	10.70	10.50	14.20	13.50
12.4.1	Household insurances - structural, contents and appliances	5.20	4.30	4.50	3.70	3.90	3.50	4.40
12.4.2	Medical insurance premiums	2.40	1.70	1.50	0.90	0.50	0.70	1.30
12.4.3	Vehicle insurance including boat insurance	9.00	8.00	7.80	6.10	5.90	9.70	7.60
12.4.4	Non-package holiday, other travel insurance	[0.20]	[0.20]	0.10	..	[0.30]	..	0.20
12.5	Other services n.e.c	5.80	4.70	4.40	2.80	3.60	2.90	4.20
12.5.1	Moving house	3.70	2.90	2.50	1.40	1.50	0.80	2.30
12.5.2	Bank, building society, post office, credit card charges	0.40	0.40	0.40	0.30	0.40	0.50	0.40
12.5.3	Other services and professional fees	1.60	1.40	1.50	1.10	1.60	1.60	1.50
1-12	**All expenditure groups**	**397.60**	**323.50**	**348.40**	**291.50**	**313.30**	**335.80**	**342.10**

Note: The commodity and service categories are not comparable with those in publications before 2001-02
The numbering system is sequential, it does not use actual COICOP codes

5.3 Detailed household expenditure by UK Countries and Government Office Regions (cont.)

2001-02 – 2002-03

based on weighted data and including children's expenditure

		North East	North West	Yorks & the Humber	East Midlands	West Midlands	East	London
Commodity or service		Average weekly household expenditure (£)						
13	**Other expenditure items**	**44.80**	**48.40**	**46.00**	**54.50**	**49.30**	**59.70**	**80.70**
13.1	Housing: mortgage interest payments, water, council tax etc.	32.70	34.40	29.40	37.90	34.90	42.80	49.00
13.2	Licences, fines and transfers	1.90	2.40	2.20	3.10	2.70	3.30	..
13.3	Holiday spending	[3.60]	4.30	4.90	6.40	4.20	5.90	11.50
13.4	Money transfers and credit	6.40	7.40	9.50	6.90	7.50	7.70	12.00
13.4.1	Money, cash gifts given to children	0.10	0.10	0.30	0.20	0.20	0.10	0.20
13.4.2	Cash gifts and donations	5.20	5.70	8.00	5.30	6.20	5.80	9.80
13.4.3	Club instalment payments (child) and interest on credit cards	1.10	1.50	1.20	1.40	1.10	1.80	2.10
Total expenditure		**340.80**	**376.00**	**351.00**	**387.50**	**362.20**	**422.60**	**485.60**
14	**Other items recorded**							
14.1	Life assurance, contributions to pension funds	16.50	18.50	26.40	22.00	19.40	24.50	26.50
14.2	Other insurance inc. Friendly Societies	0.50	0.80	1.20	1.40	0.90	1.10	1.10
14.3	Income tax, payments less refunds	53.40	56.80	56.00	71.40	61.00	84.40	118.10
14.4	National insurance contributions	16.90	16.80	15.60	19.50	18.70	22.10	25.30
14.5	Purchase or alteration of dwellings, mortgages	26.60	24.10	21.70	35.30	23.00	..	..
14.6	Savings and investments	5.10	5.80	4.50	9.60	6.00	10.00	7.10
14.7	Pay off loan to clear other debt	2.90	2.40	3.20	3.00	2.40	2.70	2.50
14.8	Windfall receipts from gambling etc.	..	1.80	3.20	..	0.90	1.40	1.10

Note: The commodity and service categories are not comparable with those in publications before 2001-02
The numbering system is sequential, it does not use actual COICOP codes

5.3 Detailed household expenditure by UK Countries and Government Office Regions (cont.) 2001-02 – 2002-03

based on weighted data and including children's expenditure

		South East	South West	England	Wales	Scotland	Northern Ireland	United Kingdom
Commodity or service		Average weekly household expenditure (£)						
13	**Other expenditure items**	**77.40**	**57.00**	**60.00**	**44.20**	**51.80**	**47.00**	**58.20**
13.1	Housing: mortgage interest payments, water, council tax etc.	51.50	39.10	40.40	27.70	35.10	25.90	38.90
13.2	Licences, fines and transfers	3.30	3.10	3.60	2.30	2.20	2.50	3.40
13.3	Holiday spending	7.30	6.60	6.40	5.00	5.50	7.30	6.30
13.4	Money transfers and credit	15.20	8.20	9.60	9.30	8.90	11.30	9.60
13.4.1	Money, cash gifts given to children	0.10	0.10	0.10	..	0.20	0.30	0.20
13.4.2	Cash gifts and donations	12.90	6.70	7.80	7.30	7.40	10.20	7.80
13.4.3	Club instalment payments (child) and interest on credit cards	2.30	1.40	1.60	1.00	1.30	0.80	1.60
Total expenditure		**475.00**	**380.50**	**408.50**	**335.70**	**365.10**	**382.80**	**400.30**
14	**Other items recorded**							
14.1	Life assurance, contributions to pension funds	27.70	19.60	23.00	16.80	20.70	18.00	22.40
14.2	Other insurance inc. Friendly Societies	1.70	1.30	1.20	1.00	0.70	0.50	1.10
14.3	Income tax, payments less refunds	98.20	65.20	77.60	51.40	62.20	47.80	74.10
14.4	National insurance contributions	22.50	16.50	19.80	15.60	17.90	15.80	19.30
14.5	Purchase or alteration of dwellings, mortgages	72.50	74.30	82.80	17.90	17.80	18.40	72.20
14.6	Savings and investments	8.30	9.70	7.40	3.90	4.40	4.80	6.90
14.7	Pay off loan to clear other debt	3.20	2.20	2.70	1.70	1.80	0.50	2.50
14.8	Windfall receipts from gambling etc.	1.60	1.30	2.40	1.40	1.50	1.60	2.30

Note: The commodity and service categories are not comparable with those in publications before 2001-02
The numbering system is sequential, it does not use actual COICOP codes

5.4 Household expenditure by urban/rural areas (GB only) 2001-02 – 2002-03

based on weighted data and including children's expenditure

	London built-up area	Other metropolitan built-up areas	Other urban: population over 250K	Other urban: population 100K to 250K	Other urban: population 25K to 100K	Other urban: population 10K to 25K	Other urban: population 3K to 10K	Rural
Average number of grossed households (thousands)	3,330	3,600	2,970	2,560	3,270	2,150	1,900	4,000
Total number of households in sample (over 2 years)	1,564	1,968	1,707	1,470	1,884	1,242	1,101	2,348
Total number of persons in sample (over 2 years)	3,800	4,775	3,974	3,553	4,507	2,996	2,520	5,669
Total number of adults in sample (over 2 years)	2,802	3,505	3,018	2,615	3,368	2,236	1,973	4,291
Weighted average number of persons per household	2.5	2.4	2.3	2.4	2.4	2.4	2.3	2.4
Commodity or service	**Average weekly household expenditure (£)**							
1 Food & non-alcoholic drinks	44.70	37.40	40.80	39.80	41.20	43.80	42.30	46.30
2 Alcoholic drinks, tobacco & narcotics	10.50	11.70	11.50	11.20	11.00	11.90	10.60	11.60
3 Clothing & footwear	27.00	21.60	20.60	21.50	21.90	22.60	20.20	22.10
4 Housing[1], fuel & power	52.90	31.70	34.00	35.50	32.30	34.30	32.30	37.00
5 Household goods & services	33.80	25.00	27.30	27.50	28.00	28.80	29.60	36.90
6 Health	6.70	3.30	3.70	4.60	4.80	4.40	4.90	5.50
7 Transport	62.80	43.80	54.30	58.10	54.00	58.20	60.60	73.10
8 Communication	14.20	9.20	10.20	9.80	10.00	9.60	9.60	10.30
9 Recreation & culture	58.80	47.50	55.10	54.30	54.60	52.70	51.90	63.50
10 Education	10.00	3.50	4.70	3.60	3.10	4.10	3.70	8.20
11 Restaurants & hotels	43.40	32.00	35.10	32.00	30.30	33.40	28.90	34.80
12 Miscellaneous goods & services	39.60	26.30	29.70	30.00	27.90	31.90	29.90	37.30
1-12 All expenditure groups	404.30	293.10	327.00	327.80	319.20	335.70	324.70	386.60
13 Other expenditure items	82.30	43.10	54.00	52.50	49.30	53.60	55.60	71.60
Total expenditure	486.60	336.20	381.00	380.30	368.50	389.30	380.30	458.20
Average weekly expenditure per person (£) Total expenditure	196.80	142.50	166.70	160.70	155.70	163.20	167.50	192.90

Note: The commodity and service categories are not comparable with those in publications before 2001-02
1 Excluding mortgage interest payments, council tax and Northern Ireland rates

Chapter 6

Trends in household expenditure

Data in Chapter 6 are presented solely for the purposes of historical comparisons. All expenditure figures are shown at 2002-03 prices (adjusted using the all items RPI). From 2001-02 figures are based on COICOP broadly mapped to the 14 main categories used on the Family Expenditure Survey, the predecessor to the Expenditure and Food Survey.

- Spending on **motoring** has increased steadily from £30 a week in 1978 to £62 in 2002-03. This was equivalent to 11 per cent of total spending in 1978 and 16 per cent in 2002-03. Spending on **leisure services** has also increased sharply over this period.

- As a proportion of total household expenditure, the largest decrease is in the spend on **food and non-alcoholic drinks**, down from 24 per cent of spending in 1978 to 17 per cent in 2002-03.

- Over the period there have also been falls in the proportion spent on **tobacco**, from 3 per cent to 1 per cent of total expenditure, on **fuel and power**, from 6 per cent to 3 per cent, and on **clothing and footwear**, from 8 per cent to 6 per cent.

- Total expenditure on leisure and household services has more than doubled as a proportion from nine per cent of all expenditure in 1978 to 20 per cent in 2002-03. The largest growth has been in **leisure services**, such as holidays and entertainment, from £16 a week in 1978 to £54 a week in 2002-03. **Household services**, such as telephone bills and domestic help, also grew substantially from £9 to £23 a week.

Chapter 6
Trends in household expenditure

Data in Chapter 6 are presented solely for the purposes of historical comparisons.

From 2001-02, the new Expenditure and Food Survey (EFS) introduced a new coding frame for expenditure items. The new frame is known as COICOP (Classification of Individual Consumption by Purpose) and is the internationally agreed standard classification for reporting household consumption expenditure within National Accounts. As such, it is part of a wider framework which helps ensure consistency across UK economic statistics. Prior to 2001-02, the EFS's predecessor survey (the Family Expenditure Survey) used a different coding system ('FES classification'). In order to preserve a time series, data presented in this chapter for the years up to 2000-01 are based on the previous FES classification. Data from 2001-02 are presented under the same classification headings, having been mapped as closely as possible from the new COICOP codes.

Please note that it is not possible to directly compare data in Chapter 6 with those in other chapters due to the differences in the definitions of the classification headings (for example, 'Motoring' in the FES classification includes vehicle insurance whereas the 'Transport' heading under COICOP excludes this expenditure item).

This chapter looks at trends in average household expenditure since 1978. The definitions of some categories of expenditure have changed over the period; definitions and changes to them since 1990 are outlined in **Appendices D and E** of this publication. Changes implemented before 1990 are documented in earlier reports. The results in **Table 6.1** for the period up to 1995-96 are calculated on an unweighted basis and exclude data from children's diaries, which had not then been introduced. From 1995-96, results are based on weighted data and include the spending from children's diaries. For 1995-96 itself, results are shown on both bases to illustrate the impact of this methodological change. For a more detailed explanation of the effect of weighting data on expenditure, see **Appendix F** – Differential grossing. All results in this chapter have been uprated to 2002-03 prices.

Figure 6.1 and Tables 6.1 and 6.2 show trends for broad categories of expenditure between 1978 and 2002-03. As a proportion of total expenditure, spending on leisure shows the largest increase over the period. Expenditure has increased from 10 to 19 per cent of total expenditure, up from £28 a week in 1978 (at 2002-03 prices) to around £74 a week in 2002-03. However, it is the leisure services element that shows the largest rise, up from six per cent of total spending in 1978 to around 14 per cent in 2002-03. Conversely, expenditure on food has shown the largest proportionate decrease, from 24 per cent in 1978 to 17 per cent in 2002-03. In monetary terms spending on food has declined slightly from £67 a week (2002-03 prices) in 1978 to £64 a week in 2002-03. The proportion spent on housing has fluctuated between 15 and 18 per cent of total expenditure over the same period. Spending on motoring has risen steadily from 11 per cent of total spending in the late-1970s to 16 per cent in 2002-03. There have been falls in the proportion spent on tobacco (three per cent to one per cent), fuel and power (six per cent to three per cent) and clothing and footwear (eight per cent to six per cent).

6.1 **Percentage of total expenditure on selected commodities/services, 1978 to 2002-03**

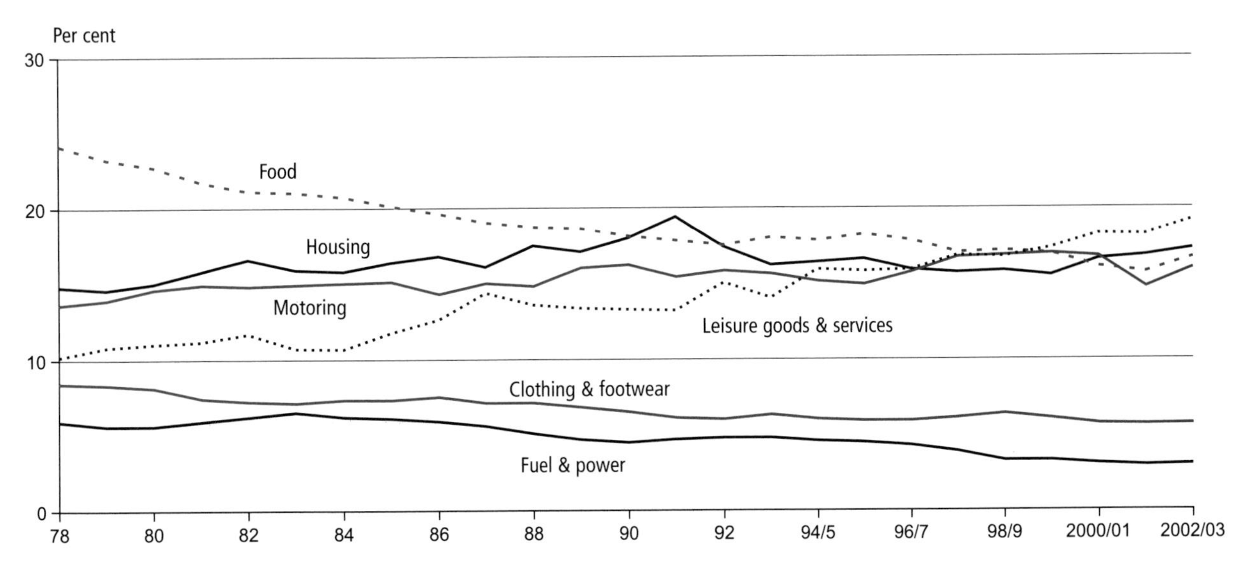

Notes:
1 There are discontinuities in housing expenditure between 1982 and 1984 and between 1991 and 1992.
2 From 1995-96 figures are based on weighted data, including children's expenditure
3 From 2001-02 data are based on COICOP expenditure codes broadly mapped to FES 14 main items

6.1 Household expenditure 1978 to 2002-03 (at 2002-03 prices)

Year		1978	1980	1982	1984	1986	1988	1990	1992	1994-95
Total number of households in sample		7,001	6,944	7,428	7,081	7,178	7,265	7,046	7,418	6,853
Total number of persons		19,019	18,844	20,022	18,557	18,330	18,280	17,437	18,174	16,617
Average number of persons per household		2.7	2.7	2.7	2.6	2.6	2.5	2.5	2.5	2.4
Commodity or service		**Average weekly household expenditure (£)**								
1	**Housing (Net)**	41.40	43.10	47.80	47.00	53.20	59.50	62.60	60.70	56.70
2	**Fuel and power**	16.60	16.00	17.90	18.40	18.60	17.40	15.60	16.70	15.80
3	**Food and non-alcoholic drinks**	67.30	65.50	60.50	61.40	62.20	63.60	63.10	61.10	61.60
4	**Alcoholic drink**	13.70	13.90	13.20	14.20	14.60	15.30	14.10	14.20	15.00
5	**Tobacco**	9.50	8.60	8.30	8.50	8.10	7.40	6.80	6.90	6.80
6	**Clothing and footwear**	23.60	23.40	20.80	21.70	24.00	24.10	22.60	21.00	20.90
7	**Household goods**	21.50	22.10	21.10	22.30	24.30	24.90	28.20	28.10	27.70
8	**Household services**	8.50	10.30	10.50	10.90	15.10	16.30	17.30	17.20	18.40
9	**Personal goods and services**	9.10	10.20	9.40	10.20	11.50	13.50	13.30	13.10	13.20
10	**Motoring**	30.20	34.10	34.60	37.20	37.80	42.00	47.60	45.70	44.20
11	**Fares and other travel costs**	7.80	7.90	7.80	7.30	7.50	8.10	8.70	9.20	8.10
12	**Leisure goods**	12.00	12.60	13.60	14.90	15.20	16.00	15.90	17.10	17.00
13	**Leisure services**	16.20	18.90	19.90	19.80	23.50	30.10	30.30	35.30	38.10
14	**Miscellaneous**	2.40	1.40	1.10	1.30	1.30	1.30	1.90	2.20	2.80
1-14	**All expenditure groups**	279.70	288.10	286.60	295.10	316.90	339.60	348.10	348.50	346.30
Average weekly expenditure per person (£)										
All expenditure groups		102.90	106.10	106.60	113.20	124.10	135.00	140.60	142.20	142.80
		Average weekly household income (£)[4]								
Gross income (£)		352.40	365.70	360.90	365.40	393.30	445.10	446.80	439.70	450.90
Disposable income (£)		287.60	298.80	291.50	292.00	319.80	361.00	363.10	359.00	364.40

6.1 Household expenditure 1978 to 2002-03 (at 2002-03 prices) (cont.)

Year		1995[1] -96	1995[2] -96	1996 -97	1997 -98	1998 -99	1999 -2000	2000 -01	2001[3] -02	2002 -03
Grossed number of households (thousands)			24,130	24,311	24,556	24,664	25,330	25,030	24,448	24,346
Total number of households in sample		6,797	6,797	6,415	6,409	6,630	7,097	6,637	7,473	6,927
Total number of persons		16,586	16,586	15,732	15,430	16,218	16,786	15,925	18,122	16,586
Weighted average number of persons per household		2.4	2.4	2.5	2.4	2.4	2.3	2.4	2.4	2.4
Commodity or service			**Average weekly household expenditure (£)**							
1	**Housing (Net)**	**57.00**	**57.50**	**56.50**	**57.50**	**61.90**	**60.90**	**66.20**	**67.20**	**66.70**
2	**Fuel and power**	**15.30**	**15.20**	**15.30**	**14.00**	**12.70**	**12.10**	**12.30**	**12.00**	**11.70**
3	**Food and non-alcoholic drinks**	**62.50**	**63.90**	**64.80**	**63.80**	**63.80**	**63.60**	**64.20**	**63.20**	**64.30**
4	**Alcoholic drink**	**13.50**	**14.50**	**15.10**	**15.80**	**15.10**	**16.30**	**15.60**	**14.60**	**14.80**
5	**Tobacco**	**6.90**	**7.00**	**7.20**	**7.10**	**6.30**	**6.40**	**6.30**	**5.60**	**5.40**
6	**Clothing and footwear**	**20.30**	**21.00**	**21.60**	**22.70**	**23.50**	**22.40**	**22.80**	**22.80**	**22.00**
7	**Household goods**	**27.70**	**28.20**	**30.70**	**30.10**	**32.10**	**32.80**	**33.80**	**33.70**	**33.80**
8	**Household services**	**17.90**	**17.90**	**18.70**	**19.80**	**20.50**	**20.20**	**22.80**	**24.10**	**23.30**
9	**Personal goods and services**	**13.70**	**13.90**	**13.60**	**14.20**	**14.40**	**14.80**	**15.20**	**15.30**	**15.20**
10	**Motoring**	**43.70**	**45.20**	**48.60**	**52.80**	**56.10**	**56.10**	**57.10**	**59.10**	**61.70**
11	**Fares and other travel costs**	**7.30**	**7.90**	**8.90**	**9.60**	**9.00**	**9.80**	**9.80**	**9.50**	**8.00**
12	**Leisure goods**	**16.20**	**17.10**	**18.20**	**19.40**	**19.30**	**19.80**	**20.50**	**20.00**	**20.50**
13	**Leisure services**	**37.90**	**38.70**	**40.40**	**44.10**	**45.40**	**46.90**	**52.40**	**53.00**	**53.60**
14	**Miscellaneous**	**2.80**	**1.40**	**1.20**	**1.20**	**1.30**	**1.50**	**0.80**	**1.90**	**1.80**
1-14	**All expenditure groups**	**342.60**	**349.30**	**361.00**	**372.00**	**381.60**	**383.60**	**399.70**	**401.90**	**402.90**
Average weekly expenditure per person (£) All expenditure groups		**140.40**	**146.00**	**144.40**	**155.00**	**159.00**	**166.80**	**169.80**	**169.90**	**169.20**
			Average weekly household income (£)[4]							
Gross income (£)		**450.20**	**460.60**	**466.40**	**479.60**	**495.10**	**512.20**	**520.70**	**551.70**	**552.30**
Disposable income (£)		**362.60**	**370.10**	**381.10**	**390.80**	**401.70**	**417.30**	**424.00**	**451.30**	**453.40**

1 From 1978 to 1995-96, figures shown are based on unweighted, adult only data.
2 From 1995-96, figures are shown based on weighted data, including children's expenditure.
3 From 2001-02 commodities and services are based on COICOP codes broadly mapped to FES.
Weighting is based on population estimates from the 2001 census.
4 Does not include imputed income from owner-occupied and rent-free households.

6.2 Household expenditure as a percentage of total expenditure 1978 to 2002-03

Year		1978	1980	1982	1984	1986	1988	1990	1992	1994
Total number of households in sample		7,001	6,944	7,428	7,081	7,178	7,265	7,046	7,418	6,853
Total number of persons		19,019	18,844	20,022	18,557	18,330	18,280	17,437	18,174	16,617
Average number of persons per household		2.7	2.7	2.7	2.6	2.6	2.5	2.5	2.5	2.4
Commodity or service		**Percentage of total expenditure**								
1	**Housing (Net)**	*15*	*15*	*17*	*16*	*17*	*18*	*18*	*17*	*16*
2	**Fuel and power**	*6*	*6*	*6*	*6*	*6*	*5*	*4*	*5*	*5*
3	**Food and non-alcoholic drinks**	*24*	*23*	*21*	*21*	*20*	*19*	*18*	*18*	*18*
4	**Alcoholic drink**	*5*	*5*	*5*	*5*	*5*	*4*	*4*	*4*	*4*
5	**Tobacco**	*3*	*3*	*3*	*3*	*3*	*2*	*2*	*2*	*2*
6	**Clothing and footwear**	*8*	*8*	*7*	*7*	*8*	*7*	*6*	*6*	*6*
7	**Household goods**	*8*	*8*	*7*	*8*	*8*	*7*	*8*	*8*	*8*
8	**Household services**	*3*	*4*	*4*	*4*	*5*	*5*	*5*	*5*	*5*
9	**Personal goods and services**	*3*	*4*	*3*	*3*	*4*	*4*	*4*	*4*	*4*
10	**Motoring**	*11*	*12*	*12*	*13*	*12*	*12*	*14*	*13*	*13*
11	**Fares and other travel costs**	*3*	*3*	*3*	*2*	*2*	*2*	*3*	*3*	*2*
12	**Leisure goods**	*4*	*4*	*5*	*5*	*5*	*5*	*5*	*5*	*5*
13	**Leisure services**	*6*	*7*	*7*	*7*	*7*	*9*	*9*	*10*	*11*
14	**Miscellaneous**	*1*	*0*	*0*	*0*	*0*	*0*	*1*	*1*	*1*
1-14	**All expenditure groups**	*100*	*100*	*100*	*100*	*100*	*100*	*100*	*100*	*100*

6.2 Household expenditure as a percentage of total expenditure 1978 to 2002-03 (cont.)

Year	1995[1] -96	1995[2] -96	1996 -97	1997 -98	1998 -99	1999 -2000	2000 -01	2001[3] -02	2002 -03
Grossed number of households (thousands)		24,130	24,310	24,560	24,660	25,330	25,030	24,450	24,350
Total number of households in sample	6,797	6,797	6,415	6,409	6,630	7,097	6,637	7,473	6,927
Total number of persons	16,586	16,586	15,732	15,430	16,218	16,786	15,925	18,122	16,586
Weighted average number of persons per household	2.4	2.4	2.5	2.4	2.4	2.3	2.4	2.4	2.4
Commodity or service		**Percentage of total expenditure**							
1 **Housing (Net)**	*17*	*16*	*16*	*15*	*16*	*16*	*17*	*17*	*17*
2 **Fuel and power**	*4*	*4*	*4*	*4*	*3*	*3*	*3*	*3*	*3*
3 **Food and non-alcoholic drinks**	*18*	*18*	*18*	*17*	*17*	*17*	*16*	*16*	*17*
4 **Alcoholic drink**	*4*	*4*	*4*	*4*	*4*	*4*	*4*	*4*	*4*
5 **Tobacco**	*2*	*2*	*2*	*2*	*2*	*2*	*2*	*1*	*1*
6 **Clothing and footwear**	*6*	*6*	*6*	*6*	*6*	*6*	*6*	*6*	*6*
7 **Household goods**	*8*	*8*	*9*	*8*	*8*	*9*	*8*	*8*	*9*
8 **Household services**	*5*	*5*	*5*	*5*	*5*	*5*	*6*	*6*	*6*
9 **Personal goods and services**	*4*	*4*	*4*	*4*	*4*	*4*	*4*	*4*	*4*
10 **Motoring**	*13*	*13*	*13*	*14*	*15*	*15*	*14*	*15*	*16*
11 **Fares and other travel costs**	*2*	*2*	*2*	*3*	*2*	*3*	*2*	*2*	*2*
12 **Leisure goods**	*5*	*5*	*5*	*5*	*5*	*5*	*5*	*5*	*5*
13 **Leisure services**	*11*	*11*	*11*	*12*	*12*	*12*	*13*	*13*	*14*
14 **Miscellaneous**	*1*	*0*	*0*	*0*	*0*	*0*	*0*	*0*	*0*
1-14 **All expenditure groups**	*100*	*100*	*100*	*100*	*100*	*100*	*100*	*100*	*100*

1 From 1978 to 1995-96, figures shown are based on unweighted, adult only data.
2 From 1995-96, figures are shown based on weighted data, including children's expenditure.
3 From 2001-02 commodities and services are based on COICOP codes broadly mapped to FES.
Weighting is based on population estimates from the 2001 census.

Chapter 7

Detailed expenditure & place of purchase

Detailed expenditures shown are averaged across all households, including those reporting zero expenditure on a specific item.

- Transport was the area of highest expenditure in 2002-03, at around £59.00 a week. Included within this category are vehicle purchases (both new and second-hand) at £26.60, operation of personal transport (including petrol and diesel costs) at £24.10 and spending on transport services such as rail and tube fares, at £8.50 a week.

- The average household spent almost £27 a week on the purchase of vehicles. Of this, £11.30 was spent on new cars or vans, £14.50 was spent on second-hand cars or vans and around £1on motorcycles.

- Spending on recreation and culture was £56 a week. Around 20 per cent of this was spent on package holidays abroad, at £11.70 a week. Of the £17.20 a week spent on recreational and cultural services, £3.70 went on gambling payments, £1.60 was spent on the cinema, theatre or museums, and £5.60 was spent on sports admissions, subscriptions and leisure class fees.

Place of purchase

- Fifty one per cent of households bought and consumed alcoholic drinks on licensed premises, spending £8.90 a week on average, of which, £5.40 went on beer and lager. Households spent an average of £5.90 a week on the purchase of alcohol from off-licences and supermarkets.

- Of the £12.70 a week spent on petrol, £3.70 was spent at large supermarket chains whilst £9.00 was spent at other outlets.

7.1 Average weekly expenditure on selected commodities and services

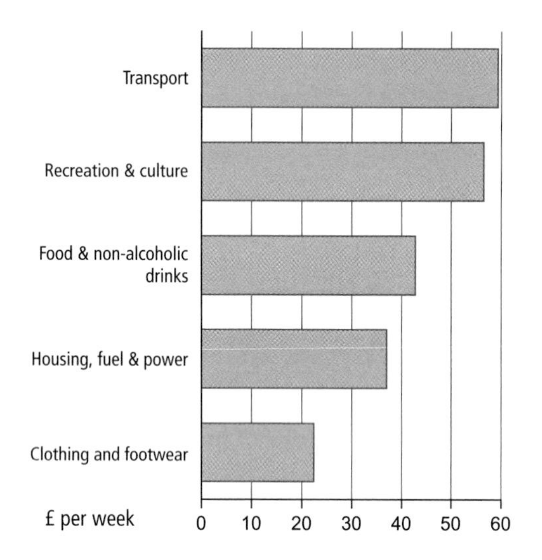

Chapter 7
Detailed expenditure and place of purchase

Detailed expenditure in 2002-03

In 2002-03, total expenditure was made up from the total of 12 COICOP expenditure groups plus 'Other expenditure items'. Other expenditure items are all items that were included in the context of Family Spending prior to the introduction of COICOP in 2001-02, but which are excluded from the COICOP expenditure groups. For a breakdown of these items see section 13 of **Table 7.1** on page 118.

Table 7.1 provides the most detailed breakdown of household expenditure. Weekly expenditure is averaged over all households in the survey and is shown for over 160 items, many of which are further sub-divided. Details concerning other items that are not classified as expenditure in the context of Family Spending are also shown, under the heading 'Other items recorded'.

The standard errors shown in **Table 7.1** take full account of the EFS sample design. More details on standard errors can be found in **Appendix C**.

Main items of expenditure

Figure 7.1 shows weekly household expenditure on a selection of goods and services. Most expenditure went on transport, at £59 a week. This was followed by recreation and culture, at £56 a week, and food and non-alcoholic drinks was the third highest item of expenditure at £43 per week. Spending on clothing and footwear was 38 per cent of that spent on transport, at around £22 per week. The lowest expenditure items were education and health, both at around £5 per week.

Transport

Expenditure on transport includes vehicle purchase, operation of personal transport and use of transport services. On average, households spent almost £27 on the purchase of vehicles. Of this, £11.30 was spent on new cars or vans, while £14.50 was spent on second-hand purchases. Spending on petrol and diesel was £14.80 a week per household, and overall, the operation of personal transport (including petrol costs, repairs, servicing and spares and accessories) constituted 41 per cent of the transport spending, at £24.10 a week.

The average weekly spend on transport services was £8.50. Rail and tube fares accounted for £1.80 of this, with a further £1.40 spent on bus and coach fares. The remainder was spent on air fares, taxis and car hire etc.

Recreation and culture

Figure 7.2 shows a breakdown of expenditure on recreation and culture. Of the £56 a week spent on this item, the largest portion went on recreational and cultural services at £17.20 a week. This included cinema, theatre and museums, at £1.60 a week, sports admissions, subscriptions and leisure class fees at £5.60 a week, and gambling payments at £3.70 a week.

Another large item of expenditure in this category was package holidays. Eighteen per cent of all households recorded expenditure on this item, the average across all households was £12.70 a week.

7.2 Expenditure on recreation and culture

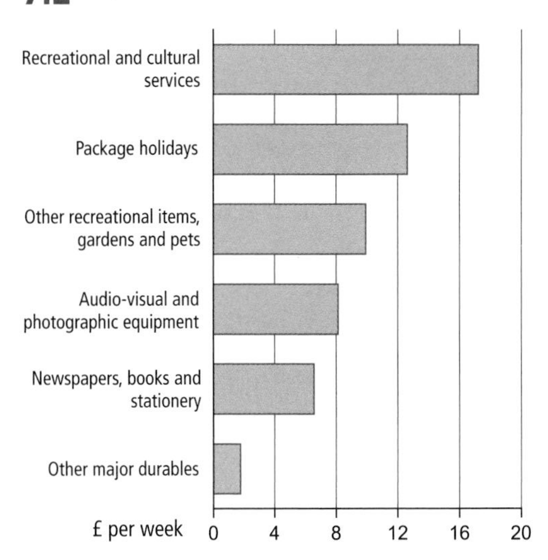

Figure 7.3 shows expenditure on the elements within other recreational items and equipment, gardens and pets. Horticultural goods, garden equipment and plants accounted for 30 per cent of the spending on this item, at £3.00 a week, while spending on pets and pet food was slightly less at £2.90 a week.

7.3 Expenditure on other recreational items and equipment, gardens and pets

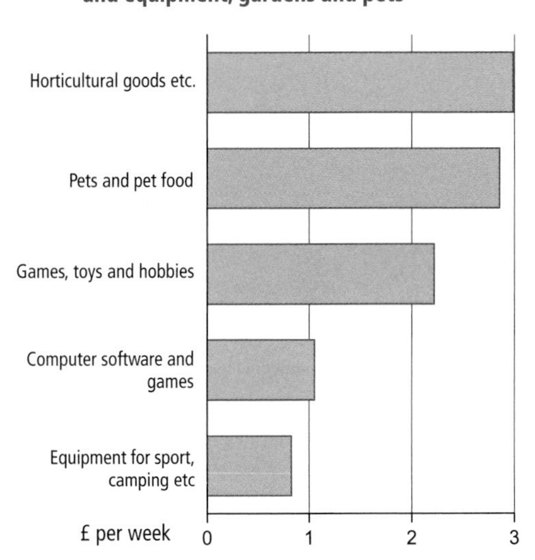

Analysis of expenditure by place of purchase

Alcoholic drinks

Table 7.2 shows expenditure on alcoholic drink by type of premises. It is split into drinks bought and consumed on licensed premises and drinks bought at off-licenses or large supermarket chains. On average, a total of £14.80 per week was spent on alcoholic drinks in 2002-03.

Fifty one per cent of households bought and consumed alcohol on licensed premises. Most of this was on beer and lager, at £5.40 a week. Of the other £5.90 spent on alcoholic drinks, households spent £4.20 a week at supermarkets and £1.80 at other off-licences.

Expenditure on food and non-alcoholic drinks

Table 7.3 gives detailed breakdown of food and non-alcoholic drinks bought from large supermarkets and separately from other outlets. Four fifths of expenditure on food and non-alcoholic drinks took place at large supermarket chains at an average of £34.10 a week, while a further £8.60 was spent at other outlets.

7.4 **Expenditure on food and non-alcoholic drinks by place of purchase**

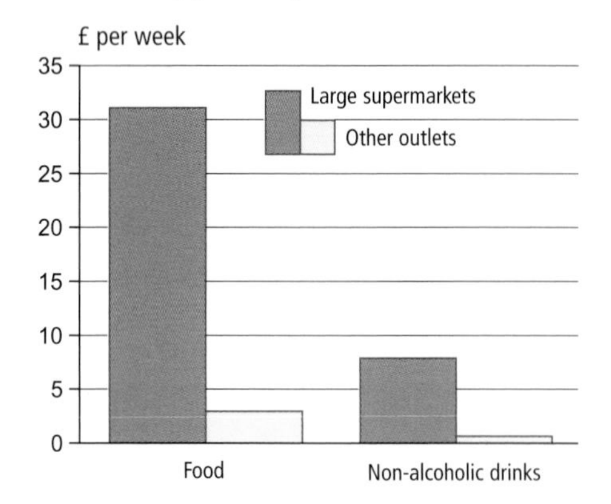

Figure 7.4 shows that for both food and non-alcoholic drinks, 80 per cent of expenditure was in large supermarket chains. £31.10 was spent on food in large supermarkets, compared to £7.90 in other outlets. The corresponding figures for non-alcoholic drinks were £3.00 and 70p respectively.

Figure 7.5 gives the expenditure on a selection of items from the food and non-alcoholic drinks category. In nearly all cases, the proportion spent in large supermarkets was markedly higher than that in other outlets. This difference can be seen most prominently for items such as fresh vegetables or cheese and curd, where 84 and 88 per cent respectively was spent at large supermarkets. The only items where similar amounts were spent in both supermarkets and other outlets were milk (43 per cent in other outlets) and lamb (40 per cent in other outlets). Only around 16 per cent of fresh vegetables were bought at other outlets.

7.5 **Expenditure on selected food items by place of purchase**

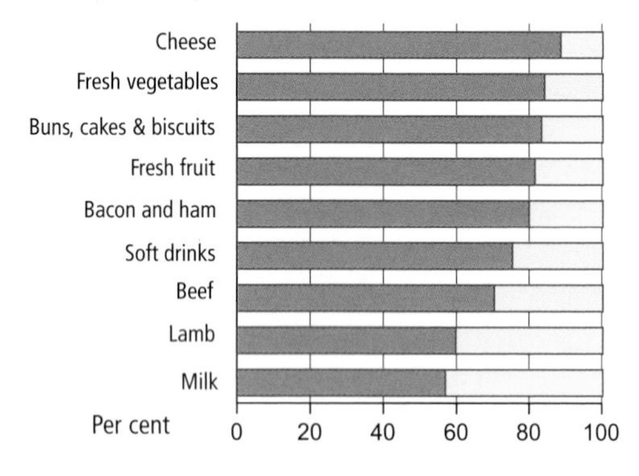

Other selected items

Table 7.4 shows expenditure on selected non-food items by place of purchase.

Petrol and diesel

Figure 7.6 shows that in the case of petrol and diesel expenditure, more was spent in other outlets than large supermarket chains. Around twice as many households bought petrol from other outlets, averaging £9 a week compared to £3.70 a week at supermarkets. Similarly, three times as many households bought diesel from other outlets, spending £1.60 a week, as opposed to 50p at supermarkets.

7.6 **Expenditure on petrol and diesel by place of purchase**

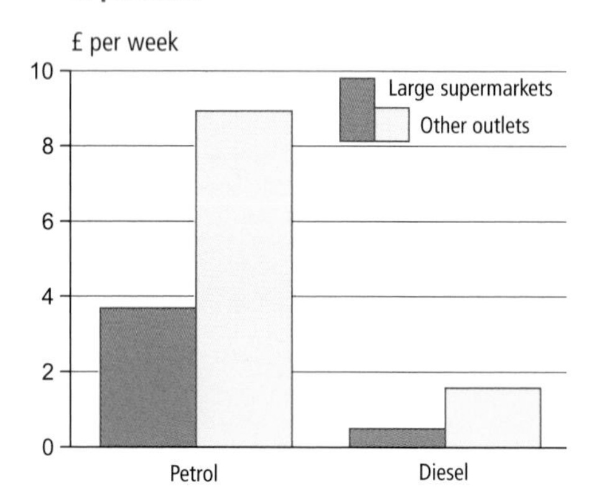

Clothing and footwear

Table 7.5 shows expenditure on clothing and footwear by place of purchase. **Figure 7.7** compares expenditure on clothing with footwear, by place of purchase. For both these items, spending was highest at other outlets, although for clothing, expenditure was spread more widely across the different places of purchase. Fifty-five per cent of clothing expenditure was at other outlets, while just over a quarter was in dedicated clothing chains, at 28 per cent, and 17 per cent was at large supermarkets. However, around four fifths of footwear expenditure was at other outlets.

7.7 Expenditure on clothing and footwear by place of purchase

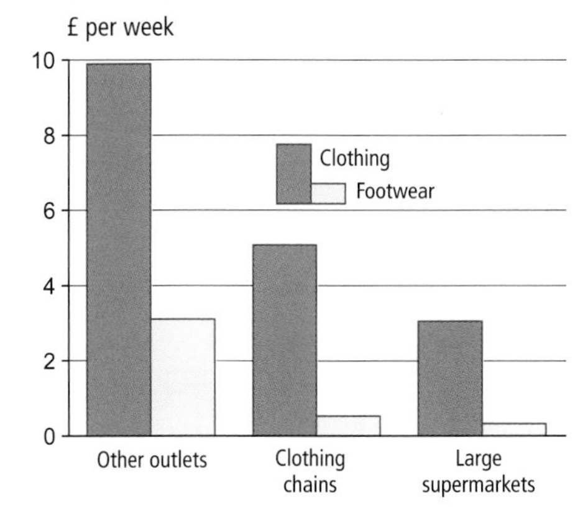

Internet Expenditure

The table below shows average weekly internet expenditure for households reporting internet purchases on the specified commodities and services. **Please note that these average figures exclude households recording zero internet expenditure.**

Figure 7.8 highlights average weekly internet spend for households ordering goods/services over the internet. Over five per cent of all households ordered goods and services over the internet in 2002-03, spending an average of £43 a week. As was the case for all households, the highest expenditure item was transport, at £79 a week. The most commonly bought items were those in the recreation and culture section, with 3.4 per cent of all households buying such items over the internet. The average weekly spend on this category, which includes books, CDs, DVDs, computer software/equipment and toys, was £24.40. This accounted for 1.5 per cent of the expenditure by all households on this category.

7.8 Average weekly internet expenditure for households ordering goods and services over the internet

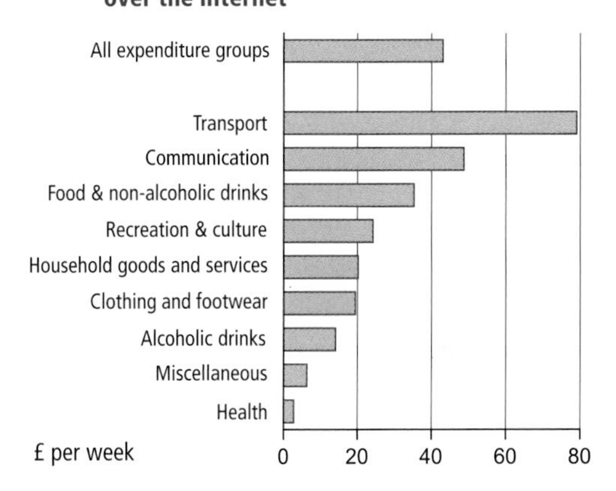

Households ordering goods/services over the internet during the two week diary period, 2002-03

Commodity or service	Recording households in sample	Grossed number of households ordering goods/services over internet (000s)	Households ordering goods/services over internet as a percentage of all households	Average weekly internet expenditure for households ordering goods/services over internet (£)	Internet expenditure as a percentage of all expenditure on the commodity or service
Food and non-alcoholic drinks	61	230	*0.9*	35.50	*0.8*
Alcoholic drinks	34	132	*0.5*	14.30	*0.7*
Clothing and footwear	51	186	*0.8*	19.60	*0.7*
Households goods and services	76	278	*1.1*	20.40	*0.8*
Health	12	42	*0.2*	3.00	*0.1*
Transport	44	180	*0.7*	79.30	*1.0*
Communication	1	4	*0.0*	48.90	*0.1*
Recreation and culture	218	832	*3.4*	24.40	*1.5*
Miscellaneous goods and services	63	234	*1.0*	6.60	*0.2*
All expenditure groups	340	1288	*5.3*	43.30	*0.7*

7.1 Components of household expenditure 2002-03

based on weighted data and including children's expenditure

	Average weekly expenditure all house-holds (£)	Total weekly expenditure (£ million)	Recording house-holds in sample	Percentage standard error (full method)
Total number of households			**6,927**	
Commodity or service				
1 Food & non-alcoholic drinks	**42.70**	**1,040**	**6,898**	***0.8***
1.1 Food	39.10	951	6,893	*0.8*
1.1.1 Bread, rice and cereals	3.90	94	6,772	*0.9*
1.1.1.1 Rice	0.20	5	1,400	*4.5*
1.1.1.2 Bread	2.00	48	6,644	*1.0*
1.1.1.3 Other breads and cereals	1.70	40	5,510	*1.2*
1.1.2 Pasta products	0.30	8	2,884	*2.5*
1.1.3 Buns, cakes, biscuits etc.	2.80	68	6,141	*1.3*
1.1.3.1 Buns, crispbread and biscuits	1.50	37	5,628	*1.4*
1.1.3.2 Cakes and puddings	1.30	31	4,629	*1.8*
1.1.4 Pastry (savoury)	0.60	15	2,415	*2.4*
1.1.5 Beef (fresh, chilled or frozen)	1.40	33	3,281	*2.1*
1.1.6 Pork (fresh, chilled or frozen)	0.60	14	1,957	*2.8*
1.1.7 Lamb (fresh, chilled or frozen)	0.60	14	1,354	*3.7*
1.1.8 Poultry (fresh, chilled or frozen)	1.50	36	3,420	*1.9*
1.1.9 Bacon and ham	0.90	21	3,490	*1.9*
1.1.10 Other meats and meat preparations	4.80	117	6,255	*1.2*
1.1.10.1 Sausages	0.60	14	3,180	*2.1*
1.1.10.2 Offal, pate etc.	0.10	3	1,134	*5.0*
1.1.10.3 Other preserved or processed meat and meat preparations	2.40	59	5,697	*1.3*
1.1.10.4 Other fresh, chilled or frozen edible meat	1.70	41	3,894	*1.9*
1.1.11 Fish and fish products	1.80	44	4,563	*1.9*
1.1.11.1 Fish (fresh, chilled or frozen)	0.70	17	2,095	*2.7*
1.1.11.2 Seafood, dried, smoked or salted fish	0.40	9	1,303	*4.2*
1.1.11.3 Other preserved or processed fish and seafood	0.70	17	3,444	*2.1*
1.1.12 Milk	2.10	52	6,477	*1.3*
1.1.12.1 Whole milk	0.70	17	2,697	*2.5*
1.1.12.2 Low fat milk	1.30	32	4,908	*1.7*
1.1.12.3 Preserved milk	0.10	3	597	*6.4*
1.1.13 Cheese and curd	1.40	34	4,956	*1.6*
1.1.14 Eggs	0.40	10	4,108	*1.6*
1.1.15 Other milk products	1.40	35	5,154	*1.5*
1.1.15.1 Other milk products	0.70	18	3,929	*1.8*
1.1.15.2 Yoghurt	0.70	16	3,621	*2.0*
1.1.16 Butter	0.30	6	2,128	*2.4*
1.1.17 Margarine and other vegetable fats	0.40	10	3,677	*1.6*
1.1.18 Peanut Butter	0.00	1	240	*7.2*
1.1.19 Cooking oils and fats	0.20	5	1,680	*3.9*
1.1.19.1 Olive oil	0.10	2	459	*6.6*
1.1.19.2 Edible oils and other edible animal fats	0.10	3	1,319	*4.7*
1.1.20 Fresh fruit	2.40	59	5,753	*1.4*
1.1.20.1 Citrus fruits (fresh)	0.40	10	3,051	*2.3*
1.1.20.2 Bananas (fresh)	0.50	12	4,317	*1.5*
1.1.20.3 Apples (fresh)	0.50	12	3,624	*1.8*
1.1.20.4 Pears (fresh)	0.10	3	1,359	*3.1*
1.1.20.5 Stone fruits (fresh)	0.30	8	1,803	*3.4*
1.1.20.6 Berries (fresh)	0.50	13	2,584	*2.3*
1.1.21 Other fresh, chilled or frozen fruits	0.20	5	1,530	*3.5*
1.1.22 Dried fruit and nuts	0.30	8	1,889	*3.4*
1.1.23 Preserved fruit and fruit based products	0.10	3	1,587	*3.4*
1.1.24 Fresh vegetables	3.00	74	6,217	*1.2*
1.1.24.1 Leaf and stem vegetables (fresh or chilled)	0.60	15	3,979	*2.2*
1.1.24.2 Cabbages (fresh or chilled)	0.40	9	3,520	*1.8*
1.1.24.3 Vegetables grown for their fruit (fresh, chilled or frozen)	1.00	25	5,074	*1.6*
1.1.24.4 Root crops, non-starchy bulbs and mushrooms (fresh, chilled or frozen)	1.00	24	5,508	*1.3*

Note: The commodity and service categories are not comparable with those in publications before 2001-02
The numbering system is sequential, it does not use actual COICOP codes

7.1 Components of household expenditure (cont.) 2002-03

based on weighted data and including children's expenditure

Commodity or service	Average weekly expenditure all households (£)	Total weekly expenditure (£ million)	Recording households in sample	Percentage standard error (full method)
1 Food & non-alcoholic drinks (continued)				
1.1.25 Dried vegetables and other preserved or processed vegetables	1.00	25	5,254	*1.8*
1.1.26 Potatoes	0.70	18	4,834	*1.4*
1.1.27 Other tubers and products of tuber vegetables	1.20	28	4,791	*1.3*
1.1.28 Sugar and sugar products	0.30	7	2,767	*2.3*
1.1.28.1 Sugar	0.20	5	2,561	*2.2*
1.1.28.2 Other sugar products	0.00	1	428	*6.8*
1.1.29 Jams, marmalades	0.20	5	2,025	*3.0*
1.1.30 Chocolate	1.20	30	4,154	*2.3*
1.1.31 Confectionery products	0.60	14	3,495	*2.0*
1.1.32 Edible ices and ice cream	0.50	12	2,321	*2.7*
1.1.33 Other food products	1.80	45	5,669	*1.5*
1.1.33.1 Sauces, condiments	0.90	22	4,252	*1.6*
1.1.33.2 Baker's yeast, dessert preparations, soups	0.70	17	4,121	*2.1*
1.1.33.3 Salt, spices, culinary herbs and other food products	0.20	5	1,555	*5.4*
1.2 Non-alcoholic drinks	3.70	89	6,348	*0.9*
1.2.1 Coffee	0.50	12	2,085	*2.5*
1.2.2 Tea	0.50	11	2,806	*2.1*
1.2.3 Cocoa and powdered chocolate	0.10	2	652	*4.4*
1.2.4 Fruit and vegetable juices, mineral waters	1.20	28	4,600	*1.6*
1.2.4.1 Fruit and vegetable juices	1.00	23	4,215	*1.7*
1.2.4.2 Mineral or spring waters	0.20	5	1,426	*3.6*
1.2.5 Soft drinks	1.50	35	4,565	*1.6*
2 Alcoholic drink, tobacco & narcotics	**11.40**	**277**	**4,529**	***1.9***
2.1 Alcoholic drinks	5.90	145	3,602	*2.4*
2.1.1 Spirits and liqueurs (brought home)	1.20	30	968	*4.6*
2.1.2 Wines, fortified wines (brought home)	2.90	70	2,390	*3.6*
2.1.2.1 Wine from grape or other fruit (brought home)	2.50	60	2,142	*3.7*
2.1.2.2 Fortified wine (brought home)	0.20	6	383	*7.2*
2.1.2.3 Champagne and sparkling wines (brought home)	0.20	4	203	*13.6*
2.1.3 Beer, lager, ciders and Perry (brought home)	1.70	41	1,907	*2.9*
2.1.3.1 Beer and lager (brought home)	1.50	38	1,732	*3.0*
2.1.3.2 Ciders and Perry (brought home)	0.20	4	366	*8.1*
2.1.4 Alcopops (brought home)	0.10	4	282	*8.9*
2.2 Tobacco and narcotics	5.40	132	2,088	*2.7*
2.2.1 Cigarettes	4.80	116	1,922	*2.9*
2.2.2 Cigars, other tobacco products and narcotics	0.60	16	510	*6.7*
2.2.2.1 Cigars	0.20	4	108	*15.3*
2.2.2.2 Other tobacco	0.40	11	408	*6.7*
2.2.2.3 Narcotics	0.00	0	7	*58.1*
3 Clothing & footwear	**22.30**	**543**	**4,808**	***2.1***
3.1 Clothing	18.30	446	4,586	*2.3*
3.1.1 Men's outer garments	4.50	109	1,458	*4.8*
3.1.2 Men's under garments	0.40	10	646	*6.1*
3.1.3 Women's outer garments	8.00	195	2,492	*3.5*
3.1.4 Women's under garments	1.30	31	1,515	*3.5*
3.1.5 Boys' outer garments (5-15)	0.80	20	554	*6.0*
3.1.6 Girls' outer garments (5-15)	1.10	27	645	*5.4*
3.1.7 Infants' outer garments (under 5)	0.60	15	581	*5.5*
3.1.8 Children's under garments (under 16)	0.40	10	711	*4.9*

Note: The commodity and service categories are not comparable with those in publications before 2001-02

7.1 Components of household expenditure (cont.) 2002-03

based on weighted data and including children's expenditure

Commodity or service	Average weekly expenditure all households (£)	Total weekly expenditure (£ million)	Recording households in sample	Percentage standard error (full method)
3 Clothing & footwear (continued)				
3.1.9 Accessories	0.70	17	982	*5.9*
3.1.9.1 Men's accessories	0.20	6	356	*8.3*
3.1.9.2 Women's accessories	0.30	8	394	*9.9*
3.1.9.3 Children's accessories	0.10	3	344	*7.2*
3.1.9.4 Protective head gear (crash helmets)	0.00	0	18	*26.1*
3.1.10 Haberdashery, clothing materials and clothing hire	0.20	5	375	*12.4*
3.1.11 Dry cleaners, laundry and dyeing	0.30	7	357	*7.6*
3.1.11.1 Dry cleaners and dyeing	0.30	6	279	*8.2*
3.1.11.2 Laundry, launderettes	0.00	1	85	*13.0*
3.2 Footwear	4.00	97	1,811	*3.3*
3.2.1 Footwear for men	1.20	28	497	*5.7*
3.2.2 Footwear for women	2.00	48	1,019	*4.5*
3.2.3 Footwear for children (5 to 15 years) and infants (under 5)	0.80	19	557	*6.5*
3.2.4 Repair and hire of footwear	0.00	1	80	*13.8*
4 Housing[1], fuel & power	**36.90**	**898**	**6,891**	***1.7***
4.1 Actual rentals for housing	21.60	526	1,967	*2.7*
4.1.1 Gross rent	21.60	526	1,960	*2.7*
4.1.2 ***less*** housing benefit, rebates and allowances received	9.40	229	1,384	*3.1*
4.1.3 Net rent	12.20	297	1,323	*4.3*
4.1.4 Second dwelling - rent	0.00	1	7	*45.9*
4.2 Maintenance and repair of dwelling	7.50	184	3,305	*5.3*
4.2.1 Central heating repairs	1.10	26	1,731	*7.4*
4.2.2 House maintenance etc.	4.10	100	1,591	*8.7*
4.2.3 Paint, wallpaper, timber	1.30	32	787	*6.3*
4.2.4 Equipment hire, small materials	1.00	25	701	*10.2*
4.3 Water supply and miscellaneous services relating to the dwelling	5.40	132	5,873	*1.8*
4.3.1 Water charges	4.60	111	5,647	*0.8*
4.3.2 Other regular housing payments including service charge for rent	0.70	18	874	*8.0*
4.3.3 Refuse collection, including skip hire	0.10	3	13	*49.5*
4.4 Electricity, gas and other fuels	11.70	285	6,551	*0.9*
4.4.1 Electricity	5.80	141	6,344	*0.9*
4.4.2 Gas	5.10	125	4,983	*1.4*
4.4.3 Other fuels	0.80	19	697	*6.6*
4.4.3.1 Coal and coke	0.20	5	189	*12.1*
4.4.3.2 Oil for central heating	0.50	13	491	*7.3*
4.4.3.3 Paraffin, wood, peat, hot water etc.	0.00	1	65	*27.5*
5 Household goods & services	**30.20**	**736**	**6,463**	***2.6***
5.1 Furniture and furnishings, carpets and other floor coverings	16.10	391	2,968	*3.7*
5.1.1 Furniture and furnishings	12.50	305	2,481	*4.2*
5.1.1.1 Furniture	11.00	267	1,625	*4.5*
5.1.1.2 Fancy, decorative goods	1.20	30	1,173	*8.6*
5.1.1.3 Garden furniture	0.30	8	63	*25.3*
5.1.2 Floor coverings	3.50	86	1,070	*6.1*
5.1.2.1 Soft floor coverings	3.10	77	1,012	*6.2*
5.1.2.2 Hard floor coverings	0.40	9	80	*26.0*
5.1.3 Repair of furniture, furnishings and floor coverings	0.00	0	02	*72.6*
5.2 Household textiles	2.10	50	1,155	*7.2*
5.2.1 Bedroom textiles, including duvets and pillows	0.80	18	446	*9.9*
5.2.2 Other household textiles, including cushions, towels, curtains	1.30	32	854	*9.5*

Note: The commodity and service categories are not comparable with those in publications before 2001-02
The numbering system is sequential, it does not use actual COICOP codes
1 Excluding mortgage interest payments, council tax and Northern Ireland rates

7.1 Components of household expenditure (cont.) 2002-03

based on weighted data and including children's expenditure

Commodity or service	Average weekly expenditure all households (£)	Total weekly expenditure (£ million)	Recording households in sample	Percentage standard error (full method)
5 Household goods & services (continued)				
5.3 Household appliances	3.20	79	790	7.6
5.3.1 Gas cookers	0.20	5	10	42.6
5.3.2 Electric cookers, combined gas/electric cookers	0.40	10	79	35.5
5.3.3 Clothes washing machines and drying machines	0.60	15	146	18.2
5.3.4 Refrigerators, freezers and fridge-freezers	0.40	9	119	19.4
5.3.5 Other major electrical appliances, dishwashers, micro-waves vacuum cleaners, heaters etc.	0.90	23	189	12.5
5.3.6 Fire extinguisher, water softener, safes etc	0.00	0	8	48.9
5.3.7 Small electric household appliances, excluding hairdryers	0.50	11	240	10.0
5.3.8 Repairs to gas and electrical appliances and spare parts	0.20	4	104	13.9
5.3.9 Rental/hire of major household appliances	0.00	1	22	30.7
5.4 Glassware, tableware and household utensils	1.60	40	1,961	5.5
5.4.1 Glassware, china, pottery, cutlery and silverware	0.70	17	834	10.6
5.4.2 Kitchen and domestic utensils	0.50	13	1,033	6.2
5.4.3 Repair of glassware, tableware and household utensils	0.00	0	1	189.0
5.4.4 Storage and other durable household articles	0.40	10	661	7.1
5.5 Tools and equipment for house and garden	2.70	65	2,449	5.6
5.5.1 Electrical tools	0.30	9	118	15.5
5.5.2 Garden tools, equipment and accessories e.g. lawn mowers etc.	0.50	11	399	10.9
5.5.3 Small tools	0.40	10	547	9.3
5.5.4 Door, electrical and other fittings	0.90	21	638	14.5
5.5.5 Electrical consumables	0.60	14	1,573	4.2
5.6 Goods and services for routine household maintenance	4.60	112	5,958	2.9
5.6.1 Cleaning materials	2.00	49	5,142	1.4
5.6.1.1 Detergents, washing-up liquid, washing powder	1.00	24	3,725	1.7
5.6.1.2 Disinfectants, polishes, other cleaning materials etc.	1.00	25	4,182	1.8
5.6.2 Household goods and hardware	1.10	26	4,251	2.3
5.6.2.1 Kitchen disposables	0.60	15	3,476	2.4
5.6.2.2 Household hardware and appliances, matches	0.20	5	917	5.5
5.6.2.3 Kitchen gloves, cloths etc.	0.10	2	1,103	4.1
5.6.2.4 Pins, needles, tape measures, nails, nuts and bolts etc.	0.10	3	496	8.1
5.6.3 Domestic services, carpet cleaning	1.50	36	1,208	7.0
5.6.3.1 Domestic services, including cleaners, gardeners, au pairs	1.10	26	363	11.1
5.6.3.2 Carpet cleaning, ironing service and window cleaner	0.40	10	940	6.1
5.6.3.3 Hire of household furniture and furnishings	0.00	0	0	
6 Health	**4.80**	**117**	**3,554**	**4.8**
6.1 Medical products, appliances and equipment	3.20	77	3,356	5.2
6.1.1 Medicines, prescriptions and healthcare products	1.50	36	3,189	2.5
6.1.1.1 NHS prescription charges and payments	0.30	7	360	7.7
6.1.1.2 Medicines and medical goods (not NHS)	1.10	26	2,907	2.5
6.1.1.3 Other medical products (e.g. plasters, condoms, hot water bottle etc.)	0.10	3	391	8.6
6.1.2 Spectacles, lenses, accessories and repairs	1.60	39	413	9.6
6.1.2.1 Purchase of spectacles, lenses, prescription sunglasses	1.60	38	343	9.8
6.1.2.2 Accessories/repairs to spectacles/lenses	0.00	1	90	13.9
6.1.3 Non-optical appliances and equipment (e.g. wheelchairs, batteries for hearing aids, shoe build-up)	0.10	2	29	35.4
6.2 Hospital services	1.70	40	560	9.6
6.2.1 Out patient services	1.50	37	552	7.3
6.2.1 NHS medical, optical, dental and medical auxiliary services	0.60	16	280	9.4
6.2.2 Private medical, optical, dental and medical auxiliary services	0.80	20	287	9.4
6.2.3 Other services	0.10	1	2	96.8
6.2.2 In-patient hospital services	0.10	3	9	79.9

Note: The commodity and service categories are not comparable with those in publications before 2001-02

7.1 Components of household expenditure (cont.) 2002-03

based on weighted data and including children's expenditure

Commodity or service	Average weekly expenditure all households (£)	Total weekly expenditure (£ million)	Recording households in sample	Percentage standard error (full method)
7 Transport	**59.20**	**1,442**	**6,066**	***2.0***
7.1 Purchase of vehicles	26.60	648	2,007	*3.7*
7.1.1 Purchase of new cars and vans	11.30	275	611	*6.1*
7.1.1.1 Outright purchases	7.00	170	231	*7.5*
7.1.1.2 Loan/Hire Purchase of new car/van	4.30	105	442	*8.0*
7.1.2 Purchase of second hand cars or vans	14.50	352	1,429	*4.5*
7.1.2.1 Outright purchases	9.80	238	878	*5.1*
7.1.2.2 Loan/Hire Purchase of second hand car/van	4.70	114	713	*6.9*
7.1.3 Purchase of motorcycles	0.90	21	134	*13.6*
7.1.3.1 Outright purchases of new or second hand motorcycles	0.40	11	55	*17.7*
7.1.3.2 Loan/Hire Purchase of new or second hand motorcycles	0.30	7	45	*23.4*
7.1.3.3 Purchase of other vehicles	0.20	4	39	*29.2*
7.2 Operation of personal transport	24.10	586	5,001	*1.7*
7.2.1 Spares and accessories	2.10	51	619	*9.1*
7.2.1.1 Car/van accessories and fittings	0.20	4	121	*18.9*
7.2.1.2 Car/van spare parts	1.70	41	409	*11.0*
7.2.1.3 Motorcycle accessories and spare parts	0.10	2	31	*27.4*
7.2.1.4 Bicycle accessories, repairs and other costs	0.20	5	110	*20.1*
7.2.2 Petrol, diesel and other motor oils	14.80	361	4,462	*1.5*
7.2.2.1 Petrol	12.70	308	4,029	*1.5*
7.2.2.2 Diesel oil	2.10	51	769	*4.3*
7.2.2.3 Other motor oils	0.10	2	120	*12.3*
7.2.3 Repairs and servicing	5.20	126	2,275	*2.9*
7.2.3.1 Car or van repairs, servicing and other work	5.10	124	2,253	*2.9*
7.2.3.2 Motorcycle repairs and servicing	0.10	2	41	*18.6*
7.2.4 Other motoring costs	1.90	47	2,707	*5.0*
7.2.4.1 Motoring organisation subscription (e.g. AA and RAC)	0.40	10	1,089	*6.2*
7.2.4.2 Garage rent, other costs (excluding fines), car washing etc.	0.50	13	502	*8.0*
7.2.4.3 Parking fees, tolls, and permits (excluding motoring fines)	0.60	14	1,582	*5.3*
7.2.4.4 Driving lessons	0.30	7	74	*23.3*
7.2.4.5 Anti-freeze, battery water, cleaning materials	0.10	3	351	*10.5*
7.3 Transport services	8.50	208	3,190	*3.9*
7.3.1 Rail and tube fares	1.80	44	835	*5.5*
7.3.1.1 Season tickets	0.60	15	134	*11.1*
7.3.1.2 Other than season tickets	1.20	29	758	*5.9*
7.3.2 Bus and coach fares	1.40	35	1,824	*4.3*
7.3.2.1 Season tickets	0.40	9	243	*7.5*
7.3.2.2 Other than season tickets	1.10	26	1,708	*4.6*
7.3.3 Combined fares	0.80	19	221	*11.3*
7.3.3.1 Combined fares other than season tickets	0.20	4	122	*12.7*
7.3.3.2 Combined fares season tickets	0.60	16	128	*13.4*
7.3.4 Other travel and transport	4.50	109	1,773	*6.3*
7.3.4.1 Air fares (within UK)	0.30	6	32	*20.1*
7.3.4.2 Air fares (international)	1.30	31	66	*18.1*
7.3.4.3 School travel	0.00	1	66	*16.8*
7.3.4.4 Taxis and hired cars with drivers	1.30	32	1,252	*4.4*
7.3.4.5 Other personal travel and transport services	0.30	7	313	*10.6*
7.3.4.6 Hire of self-drive cars, vans, bicycles	0.20	5	48	*21.4*
7.3.4.7 Car leasing	0.90	22	163	*11.0*
7.3.4.8 Water travel, ferries and season tickets	0.20	5	60	*28.6*

Note: The commodity and service categories are not comparable with those in publications before 2001-02
The numbering system is sequential, it does not use actual COICOP codes

7.1 Components of household expenditure (cont.) 2002-03

based on weighted data and including children's expenditure

Commodity or service		Average weekly expenditure all households (£)	Total weekly expenditure (£ million)	Recording households in sample	Percentage standard error (full method)
8	**Communication**	**10.60**	**258**	**6,652**	**1.2**
8.1	Postal services	0.50	12	1,792	4.2
8.2	Telephone and telefax equipment	0.60	15	215	9.7
8.2.1	Telephone purchase	0.10	4	77	16.5
8.2.2	Mobile phone purchase	0.50	11	133	11.5
8.2.3	Answering machine, fax machine, modem purchase	0.00	1	09	43.5
8.3	Telephone and telefax services	9.50	231	6,614	1.1
8.3.1	Telephone account	6.00	145	6,359	1.1
8.3.2	Telephone coin and other payments	0.10	4	270	14.8
8.3.3	Mobile phone account	2.20	54	1,546	3.6
8.3.4	Mobile phone - other payments	1.20	28	981	3.7
9	**Recreation & culture**	**56.40**	**1,373**	**6,872**	**2.1**
9.1	Audio-visual, photographic and information processing equipment	8.20	199	2,412	6.3
9.1.1	Audio equipment and accessories, CD players	2.30	55	1,290	5.1
9.1.1.1	Audio equipment, CD players including in car	0.80	19	208	12.1
9.1.1.2	Audio accessories e.g. tapes, headphones etc.	1.50	36	1,163	4.1
9.1.2	TV, video and computers	5.00	123	1,421	9.1
9.1.2.1	Purchase of TV and digital decoder	0.90	22	204	13.5
9.1.2.2	Satellite dish purchase and installation	0.00	1	8	49.6
9.1.2.3	Cable TV connection	0.00	0	4	80.4
9.1.2.4	Video recorder	0.60	15	117	18.3
9.1.2.5	Blank, pre-recorded video cassettes	1.10	27	858	5.3
9.1.2.6	Personal computers, printers and calculators	2.20	53	374	18.4
9.1.2.7	Spare parts for TV, video, audio	0.10	2	85	15.8
9.1.2.8	Repair of audio-visual, photographic and information processing	0.10	3	41	22.1
9.1.3	Photographic, cinematographic and optical equipment	0.90	21	365	15.3
9.1.3.1	Photographic and cinematographic equipment	0.70	18	158	18.3
9.1.3.2	Camera films	0.10	2	201	8.5
9.1.3.3	Optical instruments, binoculars, telescopes, microscopes	0.00	1	27	31.3
9.2	Other major durables for recreation and culture	1.80	43	163	29.1
9.2.1	Purchase of boats, trailers and horses	0.30	8	8	81.2
9.2.2	Purchase of caravans, mobile homes (including decoration)	0.70	16	28	61.6
9.2.3	Accessories for boats, horses, caravans and motor caravans	0.30	6	23	40.0
9.2.4	Musical instruments (purchase and hire)	0.10	3	59	21.5
9.2.5	Major durables for indoor recreation	0.10	1	7	72.3
9.2.6	Maintenance and repair of other major durables	0.20	6	38	22.3
9.2.7	Purchase of motor caravan (new and second-hand) - outright purchase	0.00	0	4	65.1
9.2.8	Purchase of motor caravan (new and second-hand) - loan/HP	0.10	2	10	40.8
9.3	Other recreational items and equipment, gardens and pets	10.00	242	4,877	3.6
9.3.1	Games, toys and hobbies	2.20	54	1,871	4.9
9.3.2	Computer software and games	1.10	26	370	8.7
9.3.2.1	Computer software and game cartridges	0.70	18	325	7.9
9.3.2.2	Computer games consoles	0.30	8	69	18.9
9.3.3	Equipment for sport, camping and open-air recreation	0.80	20	571	8.4
9.3.4	Horticultural goods, garden equipment and plants etc.	3.00	73	2,683	8.8
9.3.4.1	BBQ and swings	0.20	5	48	34.0
9.3.4.2	Plants, flowers, seeds, fertilisers, insecticides	2.70	65	2,584	9.5
9.3.4.3	Garden decorative	0.10	2	69	21.0
9.3.4.4	Artificial flowers, pot pourri	0.10	1	139	14.5
9.3.5	Pets and pet food	2.90	70	2,670	4.8
9.3.5.1	Pet food	1.50	36	2,550	3.0
9.3.5.2	Pet purchase and accessories	0.60	16	658	11.8
9.3.5.3	Veterinary and other services for pets identified separately	0.70	18	198	11.4

Note: The commodity and service categories are not comparable with those in publications before 2001-02

7.1 Components of household expenditure (cont.) 2002-03

based on weighted data and including children's expenditure

Commodity or service	Average weekly expenditure all households (£)	Total weekly expenditure (£ million)	Recording households in sample	Percentage standard error (full method)
9 Recreation & culture (continued)				
9.4 Recreational and cultural services	17.20	420	6,550	*2.3*
9.4.1 Sports admissions, subscriptions and leisure class fees	5.60	135	2,878	*6.0*
9.4.1.1 Spectator sports: admission charges	0.70	18	216	*34.2*
9.4.1.2 Participant sports (excluding subscriptions)	1.20	30	1,364	*4.9*
9.4.1.3 Subscriptions to sports and social clubs	1.60	40	1,194	*9.8*
9.4.1.4 Leisure class fees	1.80	45	1,314	*7.0*
9.4.1.5 Hire of equipment for sport and open air recreation	0.10	2	55	*25.9*
9.4.2 Cinema, theatre and museums etc.	1.60	40	1,139	*5.5*
9.4.2.1 Cinemas	0.50	13	647	*4.6*
9.4.2.2 Live entertainment: theatre, concerts, shows	0.80	20	364	*9.8*
9.4.2.3 Museums, zoological gardens, theme parks, houses and gardens	0.30	8	271	*8.8*
9.4.3 TV, video, satellite rental, cable subscriptions, TV licences and Internet	4.80	118	6,071	*1.1*
9.4.3.1 TV licences	1.80	44	5,849	*0.6*
9.4.3.2 Satellite subscriptions	1.60	38	1,567	*2.7*
9.4.3.3 Rent for TV/Satellite/VCR	0.40	9	463	*6.0*
9.4.3.4 Cable subscriptions	0.80	19	831	*5.1*
9.4.3.5 TV slot meter payments	0.00	1	215	*28.4*
9.4.3.6 Video, cassette and CD hire	0.10	4	425	*6.6*
9.4.3.7 Internet subscription fees	0.10	4	262	*7.3*
9.4.4 Miscellaneous entertainments	1.00	25	1,763	*4.4*
9.4.4.1 Admissions to clubs, dances, discos, bingo	0.60	15	973	*5.8*
9.4.4.2 Social events and gatherings	0.20	5	388	*9.8*
9.4.4.3 Subscriptions for leisure activities and other subscriptions	0.20	5	672	*7.3*
9.4.5 Development of film, deposit for film development, passport photos, holiday and school photos	0.50	11	563	*11.6*
9.4.6 Gambling payments	3.70	91	4,001	*3.5*
9.4.6.1 Football pools stakes	0.10	2	143	*15.3*
9.4.6.2 Bingo stakes excluding admission	0.50	11	326	*10.3*
9.4.6.3 Lottery	2.30	56	3,578	*2.3*
9.4.6.4 Bookmaker, tote, other betting stakes	0.90	22	1,166	*10.3*
9.5 Newspapers, books and stationery	6.60	160	6,360	*1.7*
9.5.1 Books, diaries, address books, cards etc.	3.60	87	4,433	*2.7*
9.5.1.1 Books	1.60	38	1,604	*4.3*
9.5.1.2 Stationery, diaries, address books, art materials	0.70	17	1,895	*4.6*
9.5.1.3 Cards, calendars, posters and other printed matter	1.30	31	3,403	*3.5*
9.5.2 Newspapers	1.90	47	5,046	*2.0*
9.5.3 Magazines and periodicals	1.10	27	3,516	*2.3*
9.6 Package holidays	12.70	308	1,217	*4.3*
9.6.1 Package holidays - UK	0.90	22	222	*9.8*
9.6.2 Package holidays - abroad	11.70	286	1,024	*4.4*
10 Education	**5.20**	**127**	**729**	***7.5***
10.1 Education fees	4.90	120	558	*7.9*
10.1.1 Nursery and primary education	1.50	37	214	*9.9*
10.1.2 Secondary education	1.00	25	57	*21.0*
10.1.3 Sixth form college/college education	0.30	6	69	*25.2*
10.1.4 University education	1.60	40	164	*11.6*
10.1.5 Other education	0.50	11	98	*21.4*
10.2 Payments for school trips, other ad-hoc expenditure	0.30	7	207	*14.0*
10.2.1 Nursery and primary education	0.10	3	115	*18.0*
10.2.2 Secondary education	0.10	3	76	*21.6*
10.2.3 Sixth form college/college education	0.00	0	3	*86.7*
10.2.4 University education	0.00	0	7	*48.0*
10.2.5 Other education	0.00	0	13	*36.8*

Note: The commodity and service categories are not comparable with those in publications before 2001-02
The numbering system is sequential, it does not use actual COICOP codes

7.1 Components of household expenditure (cont.) 2002-03

based on weighted data and including children's expenditure

Commodity or service	Average weekly expenditure all households (£)	Total weekly expenditure (£ million)	Recording households in sample	Percentage standard error (full method)
11 Restaurants & hotels	**35.40**	**863**	**6,216**	***1.8***
11.1 Catering services	30.50	742	6,182	*1.9*
11.1.1 Restaurant and café meals	11.30	275	5,093	*2.1*
11.1.2 Alcoholic drinks (away from home)	8.90	216	3,515	*2.6*
11.1.3 Take away meals eaten at home	3.60	87	3,261	*2.1*
11.1.4 Other take-away and snack food	4.20	103	4,267	*1.7*
11.1.4.1 Hot and cold food	2.80	68	3,692	*1.9*
11.1.4.2 Confectionery	0.50	11	2,698	*2.4*
11.1.4.3 Ice cream	0.10	3	888	*4.8*
11.1.4.4 Soft drinks	0.80	20	2,898	*2.1*
11.1.5 Contract catering (food)	0.70	17	54	*41.6*
11.1.6 Canteens	1.80	45	2,370	*2.8*
11.1.6.1 School meals	0.50	13	801	*4.8*
11.1.6.2 Meals bought and eaten at the workplace	1.30	32	1,932	*3.4*
11.2 Accommodation services	5.00	121	1,210	*4.7*
11.2.1 Holiday in the UK	2.50	61	848	*4.5*
11.2.2 Holiday abroad	2.40	59	415	*8.5*
11.2.3 Room hire	0.00	1	22	*39.8*
12 Miscellaneous goods and services	**33.10**	**806**	**6,782**	***1.8***
12.1 Personal care	8.70	212	6,083	*1.6*
12.1.1 Hairdressing, beauty treatment	2.70	66	1,915	*3.2*
12.1.2 Toilet paper	0.70	16	3,537	*1.5*
12.1.3 Toiletries and soap	1.90	46	4,742	*1.7*
12.1.3.1 Toiletries (disposable including tampons, lipsyl, toothpaste etc.)	1.10	28	3,988	*1.9*
12.1.3.2 Bar of soap, liquid soap, shower gel etc.	0.30	7	1,849	*3.0*
12.1.3.3 Toilet requisites (durable including razors, hairbrushes, toothbrushes etc	0.50	11	1,716	*3.4*
12.1.4 Baby toiletries and accessories (disposable)	0.50	13	940	*3.7*
12.1.5 Hair products, cosmetics and electrical appliances for personal care	2.90	70	3,808	*2.5*
12.1.5.1 Hair products	0.70	17	2,367	*2.5*
12.1.5.2 Cosmetics and related accessories	2.00	48	2,714	*3.2*
12.1.5.3 Electrical appliances for personal care, including hairdryers, shavers etc.	0.20	6	185	*10.2*
12.2 Personal effects	2.80	69	1,696	*8.7*
12.2.1 Jewellery, clocks and watches and other personal effects	1.70	42	1,104	*12.4*
12.2.2 Leather and travel goods (excluding baby items)	0.70	17	668	*7.7*
12.2.3 Sunglasses (non-prescription)	0.10	2	93	*24.7*
12.2.4 Baby equipment (excluding prams and pushchairs)	0.10	2	58	*19.5*
12.2.5 Prams, pram accessories and pushchairs	0.10	3	34	*32.5*
12.2.6 Repairs to personal goods	0.10	2	57	*23.5*
12.3 Social protection	2.60	64	417	*9.3*
12.3.1 Residential homes	0.20	4	14	*44.7*
12.3.2 Home help	0.40	9	83	*23.2*
12.3.3 Nursery, crèche, playschools	0.70	18	138	*17.3*
12.3.4 Child care payments	1.40	33	205	*12.8*

Note: The commodity and service categories are not comparable with those in publications before 2001-02
The numbering system is sequential, it does not use actual COICOP codes

7.1 Components of household expenditure (cont.) 2002-03

based on weighted data and including children's expenditure

Commodity or service	Average weekly expenditure all house-holds (£)	Total weekly expenditure (£ million)	Recording house-holds in sample	Percentage standard error (full method)
12 Miscellaneous goods and services (continued)				
12.4 Insurance	14.70	357	6,108	*1.7*
12.4.1 Household insurances	4.50	109	5,635	*1.3*
12.4.1.1 Structure insurance	2.10	52	4,434	*1.4*
12.4.1.2 Contents insurance	2.20	55	5,395	*1.5*
12.4.1.3 Insurance for household appliances	0.10	3	110	*21.1*
12.4.2 Medical insurance premiums	1.40	34	727	*6.1*
12.4.3 Vehicle insurance including boat insurance	8.60	209	4,886	*2.4*
12.4.3.1 Vehicle insurance	8.60	209	4,885	*2.4*
12.4.3.2 Boat insurance (not home)	0.00	0	5	*56.7*
12.4.4 Non-package holiday, other travel insurance	0.20	5	64	*19.8*
12.5 Other services	4.30	104	2,969	*5.9*
12.5.1 Moving house	2.40	60	591	*8.0*
12.5.1.1 Moving and storage of furniture	0.30	7	243	*11.1*
12.5.1.2 Property transaction - purchase and sale	0.80	20	138	*13.9*
12.5.1.3 Property transaction - sale only	0.70	17	111	*15.2*
12.5.1.4 Property transaction - purchase only	0.50	13	217	*15.4*
12.5.1.5 Property transaction - other payments	0.10	3	124	*11.6*
12.5.2 Bank, building society, post office, credit card charges	0.40	9	1,560	*4.0*
12.4.5.1 Bank and building society charges	0.30	6	955	*5.1*
12.4.5.2 Bank and Post Office counter charges	0.00	0	20	*28.6*
12.4.5.3 Annual standing charge for credit cards	0.00	1	616	*6.9*
12.4.5.4 Commission travellers' cheques and currency	0.10	1	160	*9.2*
12.5.3 Other services and professional fees	1.40	35	1570	*9.6*
12.5.3.1 Other professional fees including court fines	0.20	5	38	*29.0*
12.5.3.2 Legal fees	0.30	8	40	*30.1*
12.5.3.3 Funeral expenses	0.00	1	10	*56.0*
12.5.3.4 TU and professional organisations	0.70	16	1278	*8.2*
12.5.3.5 Other payments for services e.g. photocopying	0.20	5	328	*20.9*
1-12 All expenditure groups	**348.30**	**8,479**	**6,927**	***1.0***
13 Other expenditure items	**57.90**	**1,410**	**6,461**	***1.6***
13.1 Housing: mortgage interest payments, water, council tax etc.	39.40	960	5,931	*1.5*
13.1.1 Mortgage interest payments	24.50	597	2,830	*2.1*
13.1.2 Mortgage protection premiums	1.40	33	1,406	*3.6*
13.1.3 Council tax, domestic rates	13.20	321	5,856	*0.8*
13.1.5 Council tax, mortgage, insurance (secondary dwelling)	0.40	9	46	*21.8*
13.2 Licences, fines and transfers	2.70	66	4,800	*2.6*
13.2.1 Stamp duty, licences and fines (excluding motoring fines)	0.30	6	139	*23.9*
13.2.2 Motoring fines	0.00	1	14	*28.7*
13.2.3 Motor vehicle road taxation payments less refunds	2.40	59	4,779	*1.2*
13.3 Holiday spending	6.40	156	302	*8.9*
13.3.1 Money spent abroad	6.40	155	298	*8.9*
13.3.2 Duty free goods bought in UK	0.00	1	7	*46.4*

Note: The commodity and service categories are not comparable with those in publications before 2001-02
The numbering system is sequential, it does not use actual COICOP codes

7.1 Components of household expenditure (cont.) 2002-03

based on weighted data and including children's expenditure

Commodity or service	Average weekly expenditure all households (£)	Total weekly expenditure (£ million)	Recording households in sample	Percentage standard error (full method)
13 Other expenditure items (continued)				
13.4 Money transfers and credit	9.40	228	4,176	*4.7*
13.4.1 Money, cash gifts given to children	0.20	4	185	*36.5*
13.4.1.1 Money given to children for specific purposes	0.20	4	180	*37.7*
13.4.1.2 Cash gifts to children (no specific purpose)	0.00	0	9	*47.8*
13.4.2 Cash gifts and donations	7.50	184	3,299	*5.8*
13.4.2.1 Money/presents given to those outside the household	3.30	80	1,418	*11.3*
13.4.2.2 Charitable donations and subscriptions	1.70	41	2,266	*5.4*
13.4.2.3 Money sent abroad	0.90	21	363	*16.8*
13.4.2.4 Maintenance allowance expenditure	1.70	42	200	*9.6*
13.4.3 Club instalment payments (child) and interest on credit cards	1.60	40	1,646	*4.2*
13.4.3.1 Club instalment payment	0.00	0	3	*89.0*
13.4.3.2 Interest on credit cards	1.60	40	1,644	*4.2*
Total expenditure	**406.20**	**9,889**	**6,927**	***0.9***
14 Other items recorded				
14.1 Life assurance, contributions to pension funds	23.00	560	4,305	*7.5*
14.1.1 Life assurance premiums eg mortgage endowment policies	7.90	192	3,470	*3.5*
14.1.2 Contributions to pension and superannuation funds etc.	8.60	210	2,165	*2.5*
14.1.3 Personal pensions	6.50	158	1,046	*25.4*
14.2 Other insurance including Friendly Societies	1.20	30	1,899	*4.6*
14.3 Income tax, payments less refunds	73.70	1,795	5,502	*3.1*
14.3.1 Income tax paid by employees under PAYE	55.60	1,355	3,643	*2.4*
14.3.2 Income tax paid direct eg by retired or unoccupied persons	2.20	54	274	*23.7*
14.3.3 Income tax paid direct by self-employed	5.80	140	364	*12.4*
14.3.4 Income tax deducted at source from income under covenant from investments or from annuities and pensions	8.40	204	3,807	*4.2*
14.3.5 Income tax on bonus earnings	2.90	70	1,241	*16.3*
14.3.6 Income tax refunds under PAYE	0.30	6	100	*30.6*
14.3.7 Income tax refunds other than PAYE	0.90	22	577	*8.4*
14.4 National insurance contribution	19.40	472	3,631	*1.4*
14.4.1 NI contributions paid by employees	19.30	469	3,589	*1.4*
14.4.2 NI contributions paid by non-employees	0.10	2	74	*23.4*
14.5 Purchase or alteration of dwellings (contracted out), mortgages	31.90	778	2,647	*5.8*
14.5.1 Outright purchase of houses, flats etc. including deposits	0.20	4	17	*50.3*
14.5.2 Capital repayment of mortgage	11.40	277	1,708	*3.2*
14.5.3 Central heating installation	0.90	23	157	*10.1*
14.5.4 DIY improvements: Double Glazing, Kitchen Units, Sheds etc.	1.10	26	150	*15.9*
14.5.5 Home improvements - contracted out	15.70	382	1,209	*10.3*
14.5.6 Bathroom fittings	0.60	14	95	
14.5.7 Purchase of materials for Capital Improvements	0.70	16	87	*24.8*
14.5.8 Purchase of second dwelling	1.50	36	59	*38.2*
14.6 Savings and investments	6.40	155	1,399	*6.2*
14.6.1 Savings, investments (excluding AVCs)	5.40	131	1,017	*6.8*
14.6.2 Additional Voluntary Contributions	0.70	18	185	*10.7*
14.6.3 Food stamps, other food related expenditure	0.20	6	374	*9.0*
14.7 Pay off loan to clear other debt	2.50	62	450	*6.3*
14.8 Windfall receipts from gambling etc.	2.00	49	807	*10.2*

Note: The commodity and service categories are not comparable with those in publications before 2001-02
The numbering system is sequential, it does not use actual COICOP codes

7.2 Expenditure on alcoholic drink by type of premises 2002-03

based on weighted data and including children's expenditure

		Average weekly expenditure all households (£)	Total weekly expenditure (£ million)	Recording households in sample
By type of premises				
11	**Bought and consumed on licenced premises:**			
11.1.2	Alcoholic drinks (away from home)	8.90	216	3,515
	11.1.2.1 Spirits and liqueurs (away from home)	1.10	26	1,110
	11.1.2.2 Wine from grape or other fruit (away from home)	1.10	27	1,242
	11.1.2.3 Fortified wine (away from home)	0.10	1	103
	11.1.2.4 Champagne and sparkling wines (away from home)	0.10	2	112
	11.1.2.5 Ciders and Perry (away from home)	0.20	4	299
	11.1.2.6 Beer and lager (away from home)	5.40	131	2,930
	11.1.2.7 Alcopops (away from home)	0.60	14	525
	11.1.2.8 Round of drinks (away from home)	0.40	10	244
2	**Bought at off-licences (including large supermarket chains):**			
2.1	Alcoholic drinks	5.90	145	3,602
	2.1.1 Spirits and liqueurs (brought home)	1.20	30	968
	2.1.2 Wines, fortified wines (brought home)	2.90	70	2,390
	2.1.2.1 Wine from grape or other fruit (brought home)	2.50	60	2,142
	2.1.2.2 Fortified wine (brought home)	0.20	6	383
	2.1.2.3 Champagne and sparkling wines (brought home)	0.20	4	203
	2.1.3 Beer, lager, ciders and Perry (brought home)	1.70	41	1,907
	2.1.3.1 Beer and lager (brought home)	1.50	38	1,732
	2.1.3.2 Ciders and Perry (brought home)	0.20	4	366
	2.1.4 Alcopops (brought home)	0.10	4	282
2A	Bought from large supermarket chains:			
2.1A	Alcoholic drinks	4.20	101	3,013
	2.1.1A Spirits and liqueurs (brought home)	0.90	22	775
	2.1.2A Wines, fortified wines (brought home)	2.10	50	2,030
	2.1.2.1A Wine from grape or other fruit (brought home)	1.70	42	1,798
	2.1.2.2A Fortified wine (brought home)	0.20	5	335
	2.1.2.3A Champagne and sparkling wines (brought home)	0.10	3	158
	2.1.3A Beer, lager, ciders and Perry (brought home)	1.10	26	1,428
	2.1.3.1A Beer and lager (brought home)	1.00	24	1,272
	2.1.3.2A Ciders and Perry (brought home)	0.10	2	276
	2.1.4A Alcopops (brought home)	0.10	3	222
2B	Bought from other off-licence outlets:			
2.1B	Alcoholic drinks	1.80	44	1,373
	2.1.1B Spirits and liqueurs (brought home)	0.30	8	270
	2.1.2B Wines, fortified wines (brought home)	0.80	20	745
	2.1.2.1B Wine from grape or other fruit (brought home)	0.70	17	673
	2.1.2.2B Fortified wine (brought home)	0.00	1	62
	2.1.2.3B Champagne and sparkling wines (brought home)	0.10	1	51
	2.1.3B Beer, lager, ciders and Perry (brought home)	0.60	15	745
	2.1.3.1B Beer and lager (brought home)	0.60	13	686
	2.1.3.2B Ciders and Perry (brought home)	0.10	1	113
	2.1.4B Alcopops (brought home)	0.00	1	74

Note: The commodity and service categories are not comparable with those in publications before 2001-02
The numbering system is sequential, it does not use actual COICOP codes

7.3 Expenditure on food and non-alcoholic drink by place of purchase

2002-03

based on weighted data and including children's expenditure

	Large supermarket chains			Other outlets		
	Average weekly expenditure all households (£)	Total weekly expenditure (£ million)	Recording households in sample	Average weekly expenditure all households (£)	Total weekly expenditure (£ million)	Recording households in sample
1 Food and non-alcoholic drinks	**34.10**	**830**	**6,623**	**8.60**	**210**	**6,359**
1.1 Food	31.10	758	6,617	7.90	194	6,300
1.1.1 Bread, rice and cereals	3.10	76	6,369	0.70	18	3,439
1.1.1.1 Rice	0.20	4	1,220	0.10	2	209
1.1.1.2 Bread	1.50	37	6,087	0.50	12	3,058
1.1.1.3 Other breads and cereals	1.50	36	5,078	0.20	5	1,271
1.1.2 Pasta products	0.30	7	2,630	0.00	1	386
1.1.3 Buns, cakes, biscuits etc.	2.30	57	5,701	0.50	11	2,341
1.1.3.1 Buns, crispbread and biscuits	1.30	31	5,155	0.20	5	1,682
1.1.3.2 Cakes and puddings	1.00	25	4,108	0.20	6	1,367
1.1.4 Pastry (savoury)	0.60	14	2,226	0.00	1	269
1.1.5 Beef (fresh, chilled or frozen)	1.00	23	2,612	0.40	10	964
1.1.6 Pork (fresh, chilled or frozen)	0.40	10	1,548	0.20	4	496
1.1.7 Lamb (fresh, chilled or frozen)	0.40	9	920	0.20	6	470
1.1.8 Poultry (fresh, chilled or frozen)	1.20	30	2,959	0.30	6	679
1.1.9 Bacon and ham	0.70	17	2,874	0.20	4	877
1.1.10 Other meats and meat preparations	4.10	101	5,775	0.70	17	2,336
1.1.10.1 Sausages	0.40	11	2,613	0.10	3	843
1.1.10.2 Offal, pate etc.	0.10	2	938	0.00	1	235
1.1.10.3 Other preserved or processed meat and meat preparations	2.00	50	5,157	0.40	10	1,769
1.1.10.4 Other fresh, chilled or frozen meat	1.60	38	3,577	0.10	3	618
1.1.11 Fish and fish products	1.50	36	4,127	0.30	8	931
1.1.11.1 Fish (fresh, chilled or frozen)	0.50	13	1,759	0.20	4	420
1.1.11.2 Seafood, dried, smoked or salted fish	0.30	7	1,114	0.10	2	241
1.1.11.3 Other preserved or processed fish and seafood	0.60	15	3,148	0.10	2	478
1.1.12 Milk	1.20	29	5,345	0.90	22	3,380
1.1.12.1 Whole milk	0.40	9	2,073	0.30	8	1,353
1.1.12.2 Low fat milk	0.80	18	4,016	0.60	14	2,326
1.1.12.3 Preserved milk	0.10	2	518	0.00	1	105
1.1.13 Cheese and curd	1.20	30	4,530	0.20	4	817
1.1.14 Eggs	0.30	7	3,152	0.10	3	1,233
1.1.15 Other milk products	1.30	32	4,806	0.10	3	942
1.1.15.1 Other milk products	0.70	17	3,654	0.10	2	572
1.1.15.2 Yoghurt	0.60	15	3,316	0.10	1	540
1.1.16 Butter	0.20	6	1,888	0.00	1	326
1.1.17 Margarine and other vegetable fats	0.40	9	3,359	0.00	1	491
1.1.18 Peanut Butter	0.00	0	208	0.00	0	33
1.1.19 Cooking oils and fats	0.20	4	1,461	0.00	1	259
1.1.19.1 Olive oil	0.10	2	391	0.00	1	69
1.1.19.2 Edible oils and other edible aniaml fats	0.10	2	1,154	0.00	1	198
1.1.20 Fresh fruit	2.00	48	5,121	0.50	11	1,950
1.1.20.1 Citrus fruits (fresh)	0.30	8	2,496	0.10	2	848
1.1.20.2 Bananas (fresh)	0.40	10	3,662	0.10	2	1,097
1.1.20.3 Apples (fresh)	0.40	10	3,008	0.10	2	913
1.1.20.4 Pears (fresh)	0.10	3	1,083	0.00	1	321
1.1.20.5 Stone fruits (fresh)	0.30	6	1,465	0.10	2	497
1.1.20.6 Berries (fresh)	0.40	11	2,189	0.10	2	609

Note: The commodity and service categories are not comparable with those in publications before 2001-02
The numbering system is sequential, it does not use actual COICOP codes

7.3 Expenditure on food and non-alcoholic drink by place of purchase (cont.) 2002-03

based on weighted data and including children's expenditure

		Large supermarket chains			Other outlets		
		Average weekly expenditure all households (£)	Total weekly expenditure (£ million)	Recording households in sample	Average weekly expenditure all households (£)	Total weekly expenditure (£ million)	Recording households in sample
1	**Food and non-alcoholic drinks (continued)**						
1.1.21	Other fresh, chilled or frozen fruits	0.20	4	1,251	0.00	1	372
1.1.22	Dried fruit and nuts	0.30	6	1,550	0.10	2	499
1.1.23	Preserved fruit and fruit based products	0.10	3	1,400	0.00	1	250
1.1.24	Fresh vegetables	2.50	62	5,701	0.50	12	2,288
	1.1.24.1 Leaf and stem vegetables (fresh or chilled)	0.50	13	3,542	0.10	2	847
	1.1.24.2 Cabbages (fresh or chilled)	0.30	7	2,905	0.10	2	975
	1.1.24.3 Vegetables grown for their fruit (fresh, chilled or frozen)	0.90	22	4,483	0.20	04	1,306
	1.1.24.4 Root crops, non-starchy bulbs and mushrooms (fresh, chilled or frozen)	0.80	20	4,853	0.20	4	1,666
1.1.25	Dried vegetables and other preserved or processed vegetables	0.50	12	4,164	0.50	13	3,193
1.1.26	Potatoes	0.60	14	3,961	0.20	4	1,446
1.1.27	Other tubers and products of tuber vegetables	1.00	25	4,357	0.10	3	1,318
1.1.28	Sugar and sugar products	0.20	5	2,430	0.00	1	496
	1.1.28.1 Sugar	0.20	4	2,234	0.00	1	465
	1.1.28.2 Other sugar products	0.00	1	390	0.00	0	44
1.1.29	Jams, marmalades	0.20	4	1,731	0.00	1	374
1.1.30	Chocolate	0.80	20	3,211	0.40	11	2,076
1.1.31	Confectionery products	0.30	8	2,419	0.20	6	1,936
1.1.32	Edible ices and ice cream	0.40	10	1,962	0.10	2	536
1.1.33	Other food products	1.60	39	5,267	0.20	6	1,288
	1.1.33.1 Sauces, condiments	0.80	20	3,939	0.10	2	629
	1.1.33.2 Baker's yeast, dessert preparations, soups	0.60	15	3,747	0.10	2	742
	1.1.33.3 Salt, spices, culinary herbs and other food products	0.10	3	1,311	0.10	2	314
1.2	Non-alcoholic drinks	3.00	72	5,815	0.70	17	2,712
1.2.1	Coffee	0.40	10	1,831	0.10	2	313
1.2.2	Tea	0.40	10	2,406	0.10	2	508
1.2.3	Cocoa and powdered chocolate	0.10	2	584	0.00	0	76
1.2.4	Fruit and vegetable juices, mineral waters	1.00	24	4,104	0.20	4	1,093
	1.2.4.1 Fruit and vegetable juices	0.80	20	3,760	0.10	3	857
	1.2.4.2 Mineral or spring waters	0.20	4	1,183	0.00	1	352
1.2.5	Soft drinks	1.10	27	3,903	0.40	9	1,890

Note: The commodity and service categories are not comparable with those in publications before 2001-02
The numbering system is sequential, it does not use actual COICOP codes

7.4 Expenditure on selected items by place of purchase 2002-03

based on weighted data and including children's expenditure

		Large Supermarket chains			Other outlets		
		Average weekly expenditure all households (£)	Total weekly expenditure (£ million)	Recording households in sample	Average weekly expenditure all households (£)	Total weekly expenditure (£ million)	Recording households in sample
7.2.2	**Petrol, diesel & other motor oils**	**4.20**	**103**	**1,796**	**10.60**	**258**	**3,572**
	10.5.1 Petrol	3.70	90	1,619	9.00	218	3,173
	10.5.2 Diesel oil	0.50	13	214	1.60	39	630
	10.5.3 Other motor oils	0.00	0	15	0.10	2	106
5	**Household goods and services**						
5.5.5	Electrical consumable	0.20	4	621	0.40	11	1,074
5.6.1	Cleaning materials	1.60	38	4,405	0.50	11	1,778
12	**Miscellaneous goods and services**						
12.1.2	Toilet paper	0.60	14	2,945	0.10	2	771
12.1.3.1 & 12.1.3.3	Toiletries and other toilet requisites - toothpaste, deodorant, tampons, razors, hairbrushes, toothbrushes	0.80	20	3,111	0.80	19	2,348
12.1.3.2	Bar of soap, liquid soap, shower gel etc	0.20	4	1,244	0.10	3	734
12.1.5.2	Cosmetics and related accessories	0.40	9	1,229	1.60	39	1,898
2.2	**Tobacco**	**1.60**	**39**	**1,025**	**3.80**	**93**	**1,806**
2.2.1	Cigarettes	1.40	34	916	3.40	82	1,667
2.2.2	Cigars and other tobacco products	0.20	5	194	0.40	10	408
	2.2.2.1 Cigars	0.10	2	52	0.10	2	82
	2.2.2.2 Other tobacco	0.10	3	144	0.30	8	331
9	**Recreation and culture**						
9.3.5.1	Pet food	0.90	21	1,963	0.60	15	1,212
9.5.2	Newspapers	0.20	5	1,995	1.70	42	4,678
9.5.3	Magazines and periodicals	0.30	7	1,593	0.80	20	2,722
9.5.1.2 & 9.5.1.3	Stationery, diaries, address books, art materials, cards, calendars, posters and other printed matter	0.30	8	1,325	1.70	41	3,591
8.1	Postal services	0.00	0	0	0.50	12	1,792

Note: The commodity and service categories are not comparable with those in publications before 2001-02
The numbering system is sequential, it does not use actual COICOP codes

7.5 Expenditure on clothing and footwear by place of purchase

2002-03

based on weighted data and including children's expenditure

			Large supermarket chains			Clothing chains			Other outlets		
			Average weekly expenditure all households (£)	Total weekly expenditure (£ million)	Recording households in sample	Average weekly expenditure all households (£)	Total weekly expenditure (£ million)	Recording households in sample	Average weekly expenditure all households (£)	Total weekly expenditure (£ million)	Recording households in sample
3	**Clothing and footwear**		**3.40**	**82**	**1,774**	**5.60**	**136**	**1,633**	**13.00**	**316**	**4,232**
3.1	Clothing		3.00	74	1,704	5.10	123	1,561	9.90	241	3,945
	3.1.1	Clothing materials	0.00	0	2	0.00	1	4	0.10	2	73
	3.1.2	Men's outer garments	0.60	14	291	1.10	28	388	2.70	67	1,192
	3.1.3	Men's under garments	0.10	3	192	0.20	4	155	0.10	3	497
	3.1.4	Women's outer garments	1.30	31	576	2.60	63	834	4.20	101	2,087
	3.1.5	Women's under garments	0.50	12	661	0.50	13	431	0.30	7	1,032
	3.1.6	Boys' outer garments	0.10	2	116	0.10	3	77	0.60	15	459
	3.1.7	Girls' outer garments	0.10	3	133	0.20	4	102	0.80	21	549
	3.1.8	Infants' outer garments	0.10	3	160	0.10	3	102	0.40	9	426
	3.1.9	Children's under garments	0.10	3	217	0.10	2	104	0.20	5	550
	3.1.10	Accessories	0.10	3	179	0.20	4	190	0.40	9	793
		3.1.10.1 Men's accessories	0.00	1	63	0.10	2	93	0.10	3	283
		3.1.10.2 Women's accessories	0.10	1	74	0.10	2	83	0.20	4	306
		3.1.10.3 Children's accessories	0.00	0	51	0.00	0	28	0.10	2	292
	3.1.11	Haberdashery and clothing hire	0.00	0	27	0.00	0	13	0.10	3	291
3.2	Footwear		0.30	8	234	0.50	13	244	3.10	76	1,584
	3.2.1	Men's	0.10	2	57	0.10	3	54	1.00	24	440
	3.2.2	Women's	0.20	5	131	0.40	9	168	1.40	34	903
	3.2.3	Children's	0.00	1	61	0.00	1	30	0.70	18	502

Note: The commodity and service categories are not comparable with those in publications before 2001-02
The numbering system is sequential, it does not use actual COICOP codes

Household income

Income is not adjusted to take into account the different composition of households (equivalisation) as done in some other income analyses.
More detailed income information is available from the Family Resources Survey (FRS) conducted by the Department for Work and Pensions (DWP).

- Average gross income was £552 a week in 2002-03, 17 per cent higher than five years earlier after allowing for inflation.

- Retired households with other forms of income had around twice the gross weekly income of those retired households mainly dependent on state pensions. For those retired households mainly dependent on state pensions over 90 per cent of income came from social security benefits compared to around 40 per cent for those with other forms of income.

- One adult retired households, not dependent on state pension, had the highest income from investments as a proportion of total income, at 15 per cent.

- In two adult households with children, as the number of children in the household increased, the proportion of income from **wages and salaries** and **self-employment** fell and the proportion from **social security benefits** increased.

- Households with the lowest 20 per cent of incomes had a gross weekly income of £123 a week compared to £1,271 a week for those with the top 20 per cent of incomes.

- London, the South East and East of England were the only regions that recorded an average gross weekly income above the UK national average of £546 (£736, £657 and £581 a week respectively). The lowest average incomes were in Yorkshire & the Humber, Wales and Northern Ireland at £458, £451 and £448 a week respectively, around 20 per cent below the UK average.

- Average household income varied with the population of the built up area. It was highest in the London built-up area, at £733 a week, followed by rural areas at £632 a week. It was lowest in other metropolitan built-up areas at £439 a week.

8.1 Gross household income, 1970 to 2002-03 at 2002-03 prices

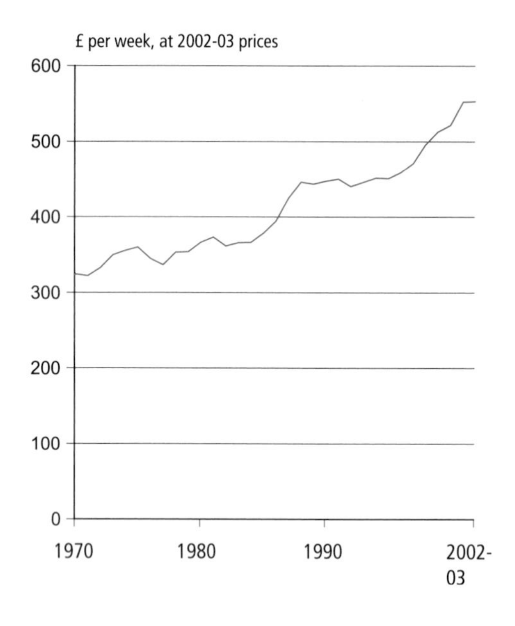

Chapter 8
Household Income

This chapter looks at the levels of gross household income, disposable income and the proportions of gross income from different sources for a variety of household types in the United Kingdom. **Income is not adjusted to take into account the different compositions of households (equivalisation) as done in some other income analyses.**

More detailed information on income is also collected in the Family Resources Survey (FRS) conducted by the Department for Work and Pensions (DWP). The FRS has the advantage of a much larger sample than the EFS, but detailed income is still required in the EFS to allow detailed analysis with expenditure. The box at the end of the chapter gives more information on the FRS and compares results from the two surveys. The FES/EFS provides a longer time series than the FRS and the 1998-99 edition of Family Spending showed changes over the last 20 to 30 years. Income does not include any Housing Benefit (HB) in this analysis. HB is treated as a reduction in housing costs.

Table 8.8 shows the changes in income since 1970 at current and at constant prices, and **Figure 8.1** shows the trends at 2002-03 prices. Average gross income was £552 a week in 2002-03, around 17 per cent higher than five years earlier after allowing for inflation.

8.2 Source of income as a percentage of gross weekly household income for all households

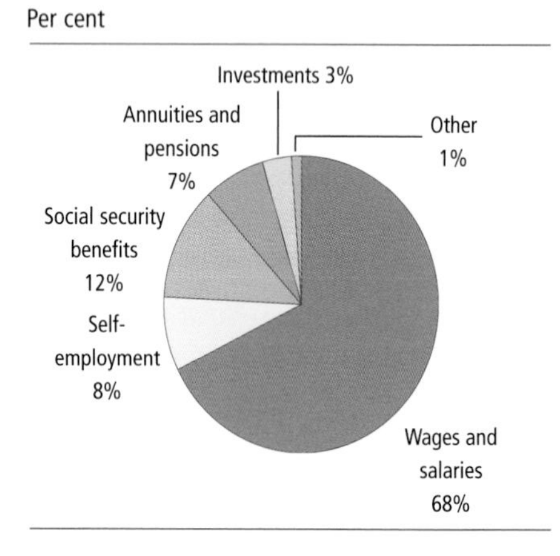

Figure 8.2 shows that in 2002-03, 68 per cent of this gross income was from wages and salaries. A further eight per cent came from self-employment and 12 per cent came from social security benefits. The remainder was split between annuities and pensions (seven per cent), investments (three per cent) and other sources (one per cent).

Household composition

Table 8.1 shows how the size and the composition of the household affected its income and sources of income in 2002-03. One-adult retired households mainly dependent on the state pension had the lowest gross income of all groups. For all households dependent on state pension, over 90 per cent of income came from social security benefits compared to around 40 per cent for retired households with other forms of income. Other retired households not dependent on state pensions had an income around twice that of dependent households with the same composition.

Households composed of a couple and one or two children had the highest proportion of income from wages and salaries and self-employment, at just over 90 per cent. The more children the couple households had, the lower the proportion of income from wages and salaries and self-employment and the higher the proportion from social security benefits (4 per cent for one child up to 19 per cent for four or more children).

Age of household reference person

Table 8.2 shows income and source of income by age of household reference person. Gross weekly income was highest for those households where the reference person was aged 30 to 49, at £695. Eighty-one per cent of this came from wages and salaries, and nine per cent from self-employment. The lowest income households were those with a reference person aged over 75, at £252 a week, 52 per cent of which came from social security benefits. Households with a reference person aged 50 to 64 had the highest percentage of income from self-employment at 11 per cent. **Figure 8.3** shows that as age of reference person increased, the proportion of income from wages and salaries and self-employment combined decreased and from social security benefits increased.

8.3 Proportion of income from wages & salaries and self-employment, and from social security benefits by age of household reference person

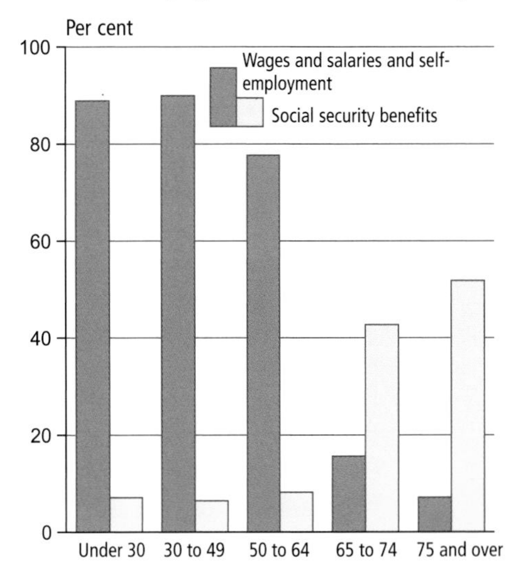

Income level

Table 8.3 highlights the average gross weekly income by gross income quintile group for 2002-03. Income ranged from £123 in the lowest quintile group to £1,271 in the highest. Social security benefits accounted for 78 per cent of income in the lowest quintile group, and this decreased steadily to just two per cent in the highest quintile group. Conversely, income from wages and salaries and self-employment combined increased from 9 per cent to 89 per cent through the quintile groups. Income from investments remained at 3 or 4 per cent for all quintile groups.

Tenure type

Table 8.4 shows that those households buying their property with a mortgage had the highest percentage of income from wages and salaries, at 82 per cent. The gross income for this group was also the highest of all tenure types, at £765 a week **(Figure 8.4)**. Social renters (either from the council or from a registered social landlord) had the highest proportion of income coming from social security benefits, at 44 per cent, and the lowest gross income, at £262, while households that owned the property outright had the highest proportion from annuities and pensions, at 22 per cent. For privately rented households, gross income was £505 a week for unfurnished properties, and £606 a week for furnished ones, whilst private renters in general received 73 per cent of their income from wages and salaries.

8.4 Gross weekly income by household tenure

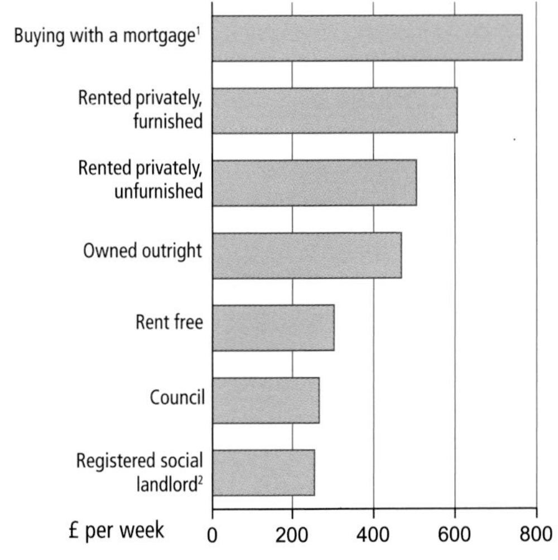

1 See footnotes in table 4.10
2 Formerly housing association

8.5 Proportion of income from wages & salaries and social security benefits by Government Office Region

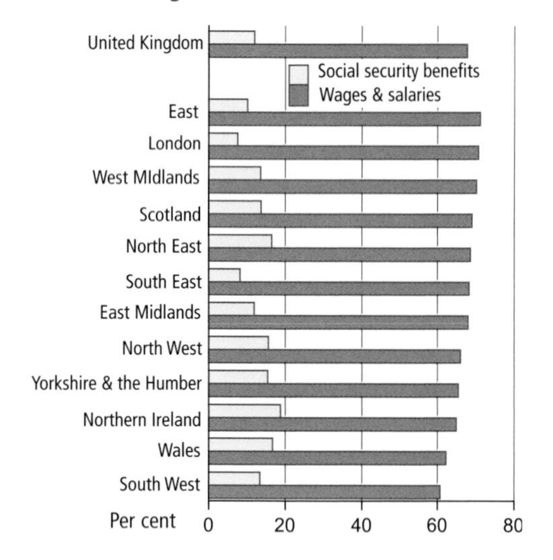

Region

Table 8.5 shows regional variations in income and source of income, averaged over two years to reduce sample variability. The average gross weekly income for the UK was £546 but this varied from £448 a week in Northern Ireland to £736 in London. Only households in the East of England, South East and London had an income above that of the UK average. **Figure 8.5** shows that for all regions, the majority of income – between 61 and 71 per cent – came from wages and salaries. The regions are ranked by the proportion of income coming from wages and salaries, and these figures are set against the proportion from social security benefits. It can be seen that as the proportion from wages and salaries drops, the proportion from social security benefits does not follow the opposite trend. Nevertheless, Northern Ireland recorded the most from social security benefits at 19 per cent, and London and the South East had the least, at 8 per cent. Households in London recorded the highest proportion of income from self-employment, at 12 per cent.

Urban/rural areas

Table 8.6 shows income and source of income by urban/rural area, averaged over the last two years (urban/rural indicators are only available for the GB sample). Households in the London built-up area had the highest gross weekly income, at £733 followed by the rural areas, at £632 a week. For all areas the main source of income was wages and salaries. Households in the London built-up area received the highest proportion from this source, at 71 per cent of gross income, closely followed by households in urban areas with a population between 100k and 250k at 70 per cent. Areas with a population between 10k and 25k showed the lowest proportion of income from wages and salaries at 64 per cent, whilst households in other metropolitan built-up areas had the lowest gross income of all areas, at £439 a week and the highest proportion of income from social security benefits, at 18 per cent.

8.6 Gross weekly income by socio-economic class

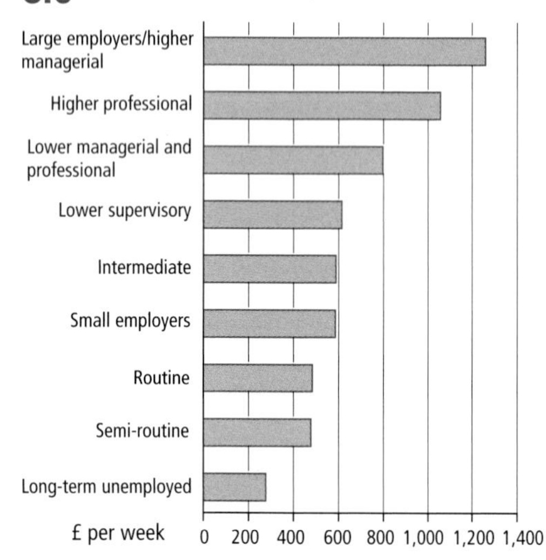

Socio-economic group

Table 8.7 illustrates gross weekly income and source of income by socio-economic group for 2002-03. Gross income varied from £1,262 a week in households where the reference person was a large employer to £279 a week where the reference person was long-term unemployed. **Figure 8.6** gives the gross weekly income for all socio-economic groups. The percentage of income from wages and salaries was highest where the reference person was in a lower supervisory role, at 92 per cent, and lowest for the small employers, whose main source of income was self-employment, at 64 per cent. Wages and salaries made up only 23 per cent of their income, the lowest of all socio-economic groups. The long-term unemployed had the highest proportion of income from social security benefits, at 17 per cent.

Family Resources Survey (FRS)

A major advantage of the Family Resources Survey is its large sample size. In the survey year 2002-03 around 27,000 GB households were interviewed for the FRS compared to just over 6,300 GB households for the Expenditure and Food Survey. The response rate was also higher than for the EFS, 63 per cent compared to 58 per cent in 2002-03.

Comparison of results from the FES and FRS

The Expenditure and Food Survey's predecessor, the Family Expenditure Survey (FES), has been compared with the FRS in the context of Households Below Average Income (HBAI) analysis. Appendix 9 of the 1979 to 1996-97 HBAI report details comparisons between the FES and the FRS. The main findings were that the FRS recorded lower equivalised disposable income, particularly for one adult and couples without children, and lower investment income, particularly for pensioners. This was due to a combination of both sampling variation and inherent differences between the two surveys. In particular it is thought that the FRS over represents some types of low-income households and under represents some types of high-income households.

Figures 8.7 and 8.8 compare the FRS and EFS results on sources of gross income as a percentage of gross weekly household income, for all GB households in 2002-03. In the FRS, 74 per cent of this income came from wages and salaries and self-employment compared to the 76 per cent recorded by the EFS. The EFS recorded a higher percentage of income from social security benefits than the FRS (12 per cent compared to nine per cent). Conversely the FRS recorded a higher percentage of income from annuities and pensions than the EFS (13 per cent compared to seven per cent). The EFS records investment income at three per cent compared to two per cent in the FRS. This supports the HBAI Appendix 9 finding that investment income was lower in the FRS.

References

Department for Work and Pensions, *Family Resources Survey Great Britain 2002-03, www.dwp.gov.uk/asd/frs.*
Department of Social Security (1977), *Households below Average Income, 1979-1996-97.* ISBN 1 84123 059 6. *www.dwp.gov.uk/asd/frs.*

8.7 EFS sources of income: all households 2002-03 Great Britain

Per cent

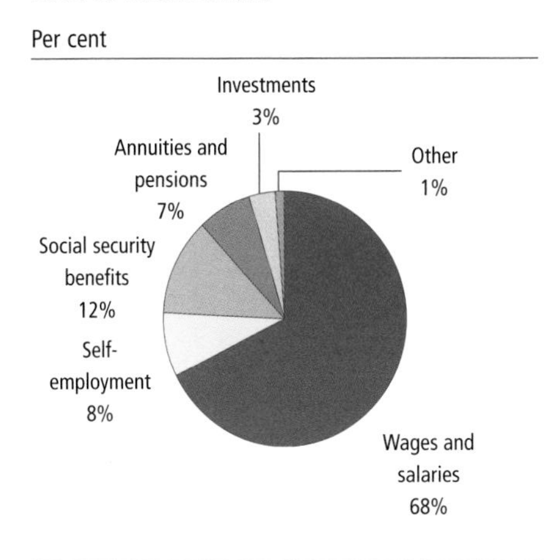

8.8 FRS sources of income: all households 2002-03 Great Britain

Per cent

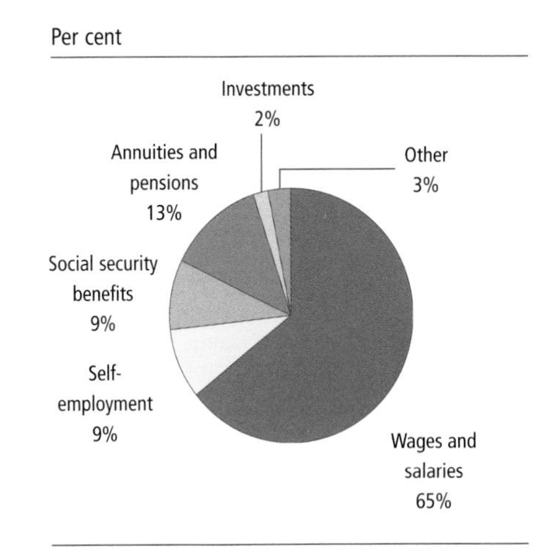

8.1 Income and source of income by household composition 2002-03

based on weighted data

	Grossed number of households	Number of households in the sample	Weekly household income: Disposable	Weekly household income: Gross	Source of income: Wages and salaries	Self employment	Investments	Annuities and pensions[1]	Social security benefits[2]	Other sources
	(000s)	Number	£	£	Percentage of gross weekly household income					
All households	**24,350**	**6,927**	**453**	**552**	***68***	***8***	***3***	***7***	***12***	***1***
Composition of household										
One adult	6,590	1,853	239	283	*51*	*6*	*6*	*13*	*23*	*1*
Retired households mainly dependent on state pensions[3]	1,640	459	121	122	*0*	*0*	*2*	*6*	*93*	*0*
Other retired households	1,470	416	228	252	*0*	*0*	*15*	*44*	*41*	*1*
Non-retired households	3,490	978	298	372	*74*	*8*	*4*	*6*	*8*	*1*
One adult, one child	700	224	231	252	*56*	*6*	*0*	*1*	*32*	*5*
One adult, two or more children	820	272	296	313	*39*	*9*	*2*	*1*	*42*	*7*
One man and one woman	7,610	2,199	471	577	*62*	*7*	*5*	*13*	*13*	*1*
Retired households mainly dependent on state pensions[3]	700	219	194	195	*0*	*0*	*2*	*7*	*92*	*0*
Other retired households	1,900	580	378	419	*4*	*1*	*12*	*44*	*38*	*1*
Non-retired households	5,020	1,399	544	689	*77*	*9*	*3*	*6*	*4*	*0*
Two men or two women	530	146	459	564	*73*	*5*	*3*	*5*	*13*	*1*
One man one woman, one child	1,740	507	607	769	*81*	*10*	*3*	*1*	*4*	*1*
One man one woman, two children	2,210	691	621	775	*82*	*10*	*1*	*1*	*5*	*1*
One man one woman, three children	660	211	595	733	*69*	*16*	*2*	*1*	*11*	*2*
Two adults, four or more children	230	75	632	776	*64*	*12*	*5*	*0*	*19*	*1*
Three adults	1,700	358	664	806	*74*	*8*	*2*	*6*	*8*	*2*
Three adults, one or more children	610	164	668	793	*78*	*9*	*2*	*1*	*9*	*2*
Four or more adults	540	119	874	1,072	*80*	*4*	*3*	*3*	*5*	*4*
Four or more adults, one or more children	270	70	786	947	*79*	*6*	*1*	*2*	*10*	*2*

1 Other than social security benefits.
2 Excluding housing benefit and council tax benefit (rates rebates in Northern Ireland) - see appendix D.
3 Mainly dependent on state pension and not economically active - see appendix D.

8.2 Income and source of income by age of household reference person 2002-03

based on weighted data

Age of household reference person	Grossed number of households	Number of households in the sample	Weekly household income: Disposable	Weekly household income: Gross	Source of income: Wages and salaries	Self employment	Investments	Annuities and pensions[1]	Social security benefits[2]	Other sources
	(000s)	Number	£	£	Percentage of gross weekly household income					
Under 30	2,550	680	427	522	*83*	*6*	*1*	*0*	*7*	*3*
30 and under 50	9,520	2,752	555	695	*81*	*9*	*2*	*1*	*6*	*1*
50 and under 65	6,040	1,682	493	606	*67*	*11*	*4*	*9*	*8*	*1*
65 and under 75	3,280	1,006	303	333	*13*	*3*	*9*	*33*	*43*	*0*
75 and over	2,960	807	234	252	*7*	*0*	*11*	*30*	*52*	*0*

1 Other than social security benefits.
2 Excluding housing benefit and council tax benefit (rates rebates in Northern Ireland) - see appendix D.

8.3 Income and source of income by gross income quintile group — 2002-03

based on weighted data

	Grossed number of house-holds	Number of house-holds in the sample	Weekly household income: Dispo-sable	Weekly household income: Gross	Source of income: Wages and salaries	Self employ-ment	Invest-ments	Annuities and pensions[1]	Social security benefits[2]	Other sources
Gross income quintile group	(000s)	Number	£	£	Percentage of gross weekly household income					
Lowest twenty per cent	4,870	1,425	121	123	*7*	*2*	*3*	*9*	*78*	*2*
Second quintile group	4,870	1,447	241	260	*30*	*5*	*4*	*16*	*43*	*2*
Third quintile group	4,870	1,416	377	437	*61*	*7*	*3*	*11*	*15*	*1*
Fourth quintile group	4,870	1,359	550	670	*76*	*7*	*3*	*7*	*6*	*1*
Highest twenty per cent	4,870	1,280	978	1,271	*79*	*10*	*4*	*4*	*2*	*1*

1 Other than social security benefits.
2 Excluding housing benefit and council tax benefit (rates rebates in Northern Ireland) - see appendix D.

8.4 Income and source of income by household tenure — 2002-03

based on weighted data

	Grossed number of house-holds	Number of house-holds in the sample	Weekly household income: Dispo-sable	Weekly household income: Gross	Source of income: Wages and salaries	Self employ-ment	Invest-ments	Annuities and pensions[1]	Social security benefits[2]	Other sources
Tenure of dwelling[3]	(000s)	Number	£	£	Percentage of gross weekly household income					
Owners										
Owned outright	7,090	2,047	397	467	*40*	*8*	*9*	*22*	*20*	*1*
Buying with a mortgage[3]	10,030	2,836	607	765	*82*	*9*	*2*	*2*	*4*	*1*
All	17,130	4,883	520	642	*69*	*9*	*4*	*8*	*9*	*1*
Social rented from										
Council	3,420	1,012	243	265	*48*	*2*	*0*	*4*	*44*	*1*
Registered social landlord [4]	1,440	400	232	254	*46*	*4*	*1*	*5*	*43*	*1*
All	4,850	1,412	240	262	*48*	*3*	*1*	*4*	*44*	*1*
Private rented										
Rent free	350	100	268	302	*57*	*8*	*3*	*6*	*25*	*2*
Rent paid, unfurnished	1,440	393	408	505	*73*	*11*	*1*	*2*	*11*	*2*
Rent paid, furnished	580	139	500	606	*78*	*7*	*4*	*0*	*3*	*8*
All	2,360	632	410	500	*73*	*9*	*2*	*2*	*10*	*4*

1 Other than social security benefits.
2 Excluding housing benefit and council tax benefit (rates rebates in Northern Ireland) - see appendix D.
3 See footnotes in Table 4.10.
4 Formerly housing association

8.5 Income and source of income by UK Countries and Government Office Regions — 2001-02 – 2002-03

based on weighted data

Government Office Regions	Average number of grossed house-holds	Total number of house-holds (over 2 years)	Weekly household income: Dispo-sable	Weekly household income: Gross	Source of income: Wages and salaries	Self employ-ment	Invest-ments	Annuities and pensions[1]	Social security benefits[2]	Other sources
	(000s)	Number	£	£	Percentage of gross weekly household income					
United Kingdom	24,400	14,400	448	546	68	8	4	7	12	1
North East	1,020	632	401	477	69	5	2	6	17	1
North West	2,780	1,599	396	473	66	7	2	8	16	1
Yorkshire and the Humber	2,130	1,159	382	458	66	6	3	8	16	1
East Midlands	1,720	974	433	528	68	9	3	7	12	1
West Midlands	2,120	1,203	404	487	70	6	3	6	14	1
East	2,230	1,278	470	581	71	7	3	7	10	1
London	2,850	1,283	587	736	71	12	4	4	8	1
South East	3,390	1,955	528	657	69	10	5	8	8	1
South West	2,160	1,282	428	514	61	9	6	10	14	1
England	20,390	11,365	459	561	68	8	4	7	12	1
Wales	1,250	712	381	451	62	6	3	11	17	1
Scotland	2,140	1,207	404	490	69	5	2	7	14	2
Northern Ireland	620	1,116	380	448	65	7	2	5	19	2

1 Other than social security benefits.
2 Excluding housing benefit and council tax benefit (rates rebates in Northern Ireland) - see appendix D.

8.6 Income and source of income by GB urban/rural area — 2001-02 – 2002-03

based on weighted data

GB urban rural areas	Average number of grossed house-holds	Total number of house-holds (over 2 years)	Weekly household income: Dispo-sable	Weekly household income: Gross	Source of income: Wages and salaries	Self employ-ment	Invest-ments	Annuities and pensions[1]	Social security benefits[2]	Other sources
	(000s)	Number	£	£	Percentage of gross weekly household income					
Urban										
London built-up area	3,330	1,564	585	733	71	11	4	5	8	1
Other metropolitan built up areas	3,600	1,968	368	439	68	5	2	6	18	1
Other urban:										
population over 250k	2,970	1,707	413	498	67	6	4	7	13	2
population 100k to 250k	2,560	1,470	414	504	70	7	3	7	13	1
population 25k to 100k	3,270	1,884	406	489	68	6	3	8	14	1
population 10k to 25k	2,150	1,242	451	545	64	9	4	9	13	1
population 3k to 10k	1,900	1,101	420	506	67	6	3	9	13	1
Rural	4,000	2,348	509	632	67	10	5	8	9	1

1 Other than social security benefits.
2 Excluding housing benefit and council tax benefit (rates rebates in Northern Ireland) - see appendix D.

8.7 Income and source of income by socio-economic class[5] 2002-03

based on weighted data

Socio-economic class	Grossed number of households (000s)	Number of households in the sample (Number)	Weekly household income: Disposable (£)	Weekly household income: Gross (£)	Wages and salaries	Self employment	Investments	Annuities and pensions[1]	Social security benefits[2]	Other sources
					Percentage of gross weekly household income					
Higher managerial and professional	2,670	707	857	1141	84	10	3	2	1	1
Large employers/higher managerial	1,070	287	937	1262	90	3	2	2	2	1
Higher professional	1,590	420	803	1060	78	14	3	3	1	1
Lower managerial and professional	4,220	1,174	632	800	87	5	3	2	2	1
Intermediate	1,440	419	483	591	86	3	2	3	6	1
Small employers	1,380	396	517	588	23	64	4	2	7	1
Lower supervisory	1,640	464	499	617	92	0	1	1	5	1
Semi-routine	1,810	512	404	480	83	2	1	3	11	1
Routine	1,640	472	407	486	86	1	1	3	9	1
Long-term unemployed[3]	340	91	263	279	45	8	1	4	17	24
Other[4]	9,200	2,692	258	281	13	1	8	29	47	2

1 Other than social security benefits.
2 Excluding housing benefit and council tax benefit (rates rebates in Northern Ireland) - see appendix D.
3 Includes those who have never worked and full-time students
4 See Appendix D for definition of NS-SEC
5 Excludes those who are economically inactive

8.8 Income and source of income 1970 to 2002-03

	Grossed number of households (000s)	Number of households in the sample (Number)	Weekly household income[1]: Current prices: Disposable (£)	Current prices: Gross (£)	2002-03 prices: Disposable (£)	2002-03 prices: Gross (£)	Wages and salaries	Self employment	Investments	Annuities and pensions[2]	Social security benefits[3]	Other sources
							Percentage of gross weekly household income					
1970		6,393	28	34	269	324	77	7	4	3	9	1
1980		6,944	115	140	299	366	75	6	3	3	13	1
1990		7,046	258	317	363	447	67	10	6	5	11	1
1995-96		6,797	307	381	363	450	64	9	5	7	14	2
1996-97		6,415	325	397	375	458	65	9	4	7	14	1
1997-98		6,409	343	421	383	470	67	8	4	7	13	1
1998-99[4]	24,660	6,630	371	457	402	495	68	8	4	7	12	1
1999-2000	25,340	7,097	391	480	417	512	66	10	5	7	12	1
2000-01	25,030	6,637	409	503	424	521	67	9	4	7	12	1
2001-02[5]	24,450	7,473	442	541	451	552	69	9	4	7	11	1
2002-03	24,350	6,927	453	552	453	552	68	8	3	7	12	1

1 Does not include imputed income from owner-occupied and rent-free households.
2 Other than social security benefits.
3 Excluding housing benefit and council tax benefit (rates rebate in Northern Ireland) and their predecessors in earlier years - see appendix D.
4 Based on weighted data from 1998-99
5 From 2001-02 onwards, weighting is based on population figures from the 2001 census

Chapter 9

Household characteristics & ownership of consumer durables

- Sixty three per cent of all households consisted of one or two persons. Only five per cent of households had five or more people in them. Around 33 per cent of households had children, more than half of these were two adults households with one or two children.

- Between 1998-99 and 2002-03 **mobile phone** ownership increased from 27 to 70 per cent. Thirty-seven per cent of households in the lowest income group owned a mobile phone compared to 88 per cent in the highest. Households in the South East reported the highest levels of mobile phone ownership; the lowest levels were reported in Northern Ireland.

- Fifty-five per cent of households had a **home computer** and 45 per cent had **internet connection** in 2002-03. The highest level of ownership for home computers and internet access were in London and the South East. Northern Ireland reported the lowest levels.

- Twenty nine per cent of households owned a **dishwasher;** this figures varied from eight per cent for households in the lowest income group to 67 per cent for the highest income households. Other large differences between income groups were reported in the ownership of **home computers** (19 per cent in the lowest group compared to 91 per cent in the highest) and **internet connection** (12 per cent in the lowest compared to 85 per cent in the highest).

- A fifth of households in the lowest income group owned a **satellite receiver**, a higher proportion than those with **home computers**, **internet access** or **dishwasher.**

- Nearly three-quarters of households in the UK owned a **car**. The lowest levels of ownership were reported in the North East and London (62 and 65 per cent respectively); the highest levels were recorded in South East, South West and East (all over 80 per cent).

- Only a quarter of households in the lowest income group owned a **car**. For the top four income groups this increased to over 90 per cent. Over half of households in the two highest income groups owned at least **two cars**.

9.1 Households by size, 2002-03

Per cent

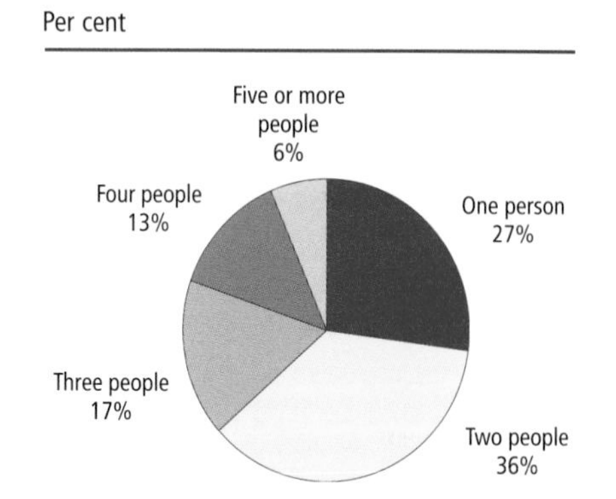

Chapter 9
Household characteristics and ownership of consumer durables

This chapter describes the characteristics of households in 2002-03 as estimated from the Expenditure and Food Survey. The relationships between ownership of durable goods and income, household composition, type of tenure and region are also outlined in this chapter.

Characteristics of households

Table 9.1 looks at characteristics of all households. The majority consisted of one or two persons at 27 and 36 per cent respectively. Only six per cent had five or more people in them (**Figure 9.1**).

Around 30 per cent of households had children. Twenty per cent were two adult households with children and a further 6 per cent of households consisted of one adult with children.

Twenty eight per cent of households had married women that were economically active, and half of these had dependent children. Of the 22 per cent of households with married but economically inactive women, only 23 per cent had dependent children, reflecting the high proportion of women above retirement age.

Seventy per cent of households were owner occupiers and the remaining 30 per cent lived in rented dwelling of which 10 per cent was privately rented.

9.2 Households with selected durable goods, 1996-97 to 2002-03

Per cent

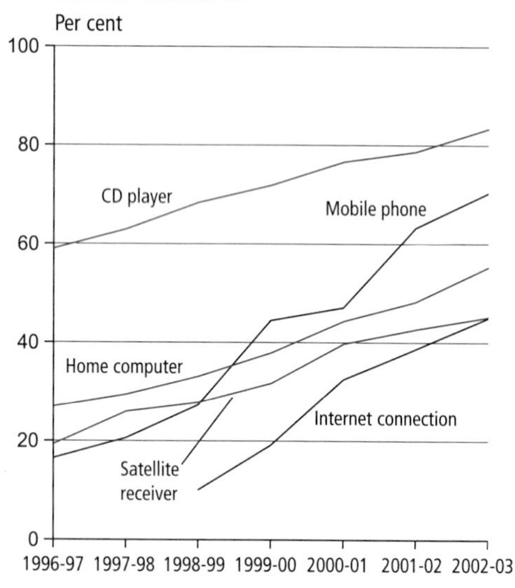

Ownership of durable goods

Table 9.3 looks at durable goods owned by households from 1970 to 2002-03. It shows that several are now almost universally owned, namely central heating, washing machine, telephone and video recorder.

Figure 9.2 illustrates changes in the proportion of households owning those goods that have undergone the largest increases since 1996-97. CD players have steadily increased and were owned by 83 per cent of households in 2002-03. Ownership of home computers and satellite receivers have also increased steadily, and in 2002-03 were owned by 45 and 55 per cent of households respectively. Households with an internet connection increased from 10 per cent in 1998-99 to 45 per cent in

2002-03, while ownership of mobile phones rose sharply, from 27 per cent in 1998-99 to 70 per cent in 2002-03.

Ownership of durable goods by income is shown in **Table 9.4** and **Figure 9.3** compares ownership levels among the highest and lowest income decile groups. The three most commonly owned goods among both groups were central heating, telephone and washing machine. Households with the top tenth of incomes had high ownership (80 to100 per cent) of most durable goods. Even the less commonly owned items such as a dishwasher were owned by 67 per cent of households in this group. In contrast, less than 10 per cent of households from the lowest income group owned a dishwasher. Other large differences between income groups were in the ownership of computers and internet access. Ninety-one per cent of households in the highest income group owned a computer with 85% having access to the internet; the corresponding figures for households in the lowest income group were 19% and 12% respectively.

9.3 Ownership of durable goods for the lowest and highest income groups, 2002-03

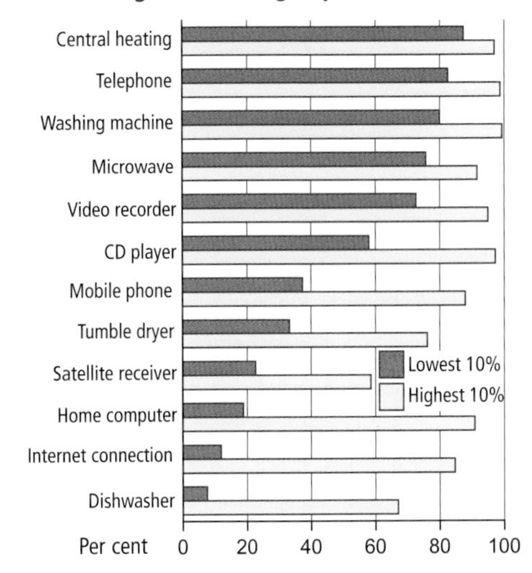

Figure 9.4 illustrates the difference in ownership of durable goods for retired and non-retired couples. The telephone, washing machine, central heating and video recorders were owned by over 80 per cent of both households groups. However, while less than 15 per cent of retired couples owned a home computer, dishwasher or had internet access, over 30 per cent of non-retired couples owned dishwashers and almost two thirds owned a home computer and had internet access. They were also twice as likely to have a mobile phone and satellite receiver.

Table 9.5 looks at car ownership by income decile group, tenure and household composition. In general, the number of households with cars increased by income group. Ninety five per cent of households in the top income groups owned a car with over two thirds owning two or more. In contrast, a quarter of households in the lowest income group owned cars with just two per cent owning two or more.

In terms of household composition, two-adult two-children households were most likely to won at least one car (94%). However, three-adult households were most likely to own at least two cars (61%).

9.4 Ownership of durable goods for retired and non-retired couple households, 2002-03

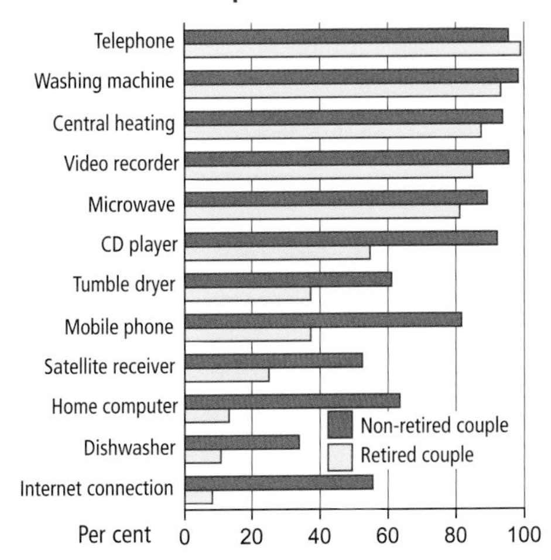

9.5 Households with cars by household tenure, 2002-03

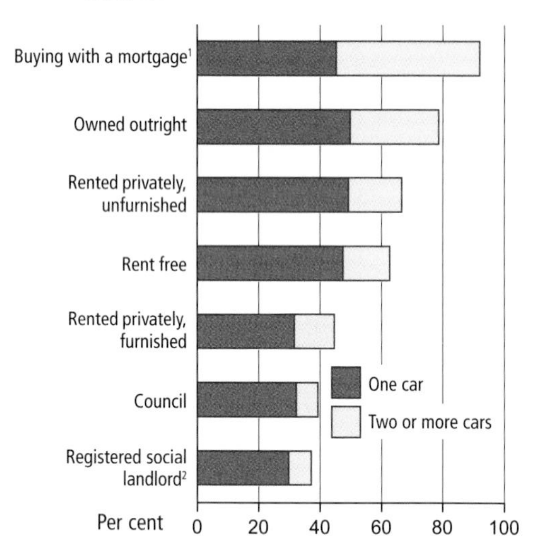

Figure 9.5 shows car ownership by household tenure. A third of households whose accommodation was rented from the council or registered social landlord owned one car with a further seven per cent owning two or more cars. Ownership was much higher among mortgaged households where 45% owned one car and a further 47% owned two or more cars.

9.6 Households with cars by region, 2002-03

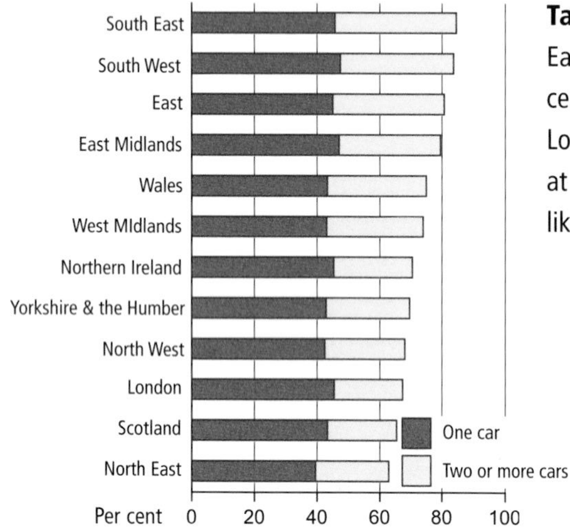

Table 9.6 and Figure 9.6 show that the South East, South West and East had the highest proportion of households with cars, all at over 80 per cent. The regions with the lowest levels of car ownership were Scotland, London and the North East where around two-thirds of households owned at least one car. Households in London and Scotland were also the least likely to own two or more cars (23 per cent).

9.1 Characteristics of households

2002-03

based on weighted data

	% * of all house-holds	Grossed number of house-holds (000s)	House-holds in sample (number)
Total number of households	*100*	24,346	6,927
Size of household			
One person	*27*	6,590	1,853
Two persons	*36*	8,850	2,570
Three persons	*17*	4,090	1,082
Four persons	*13*	3,250	953
Five persons	*4*	1,090	328
Six persons	*1*	340	104
Seven persons	*0*	90	29
Eight persons	*0*	20	6
Nine or more persons	*0*	0	2
Composition of household			
One adult	*27*	6,590	1,853
Retired households mainly dependent on state pensions[1]	*7*	1,640	459
Other retired households	*6*	1,470	416
Non-retired households	*14*	3,490	978
One man	*11*	2,800	773
Aged under 65	*8*	2,060	546
Aged 65 and over	*3*	740	227
One woman	*16*	3,790	1,080
Aged under 60	*5*	1,210	365
Aged 60 and over	*11*	2,580	715
One adult, one child	*3*	700	224
One man, one child	*0*	90	24
One woman, one child	*3*	610	200
One adult, two or more children	*3*	820	272
One man, two or more children	*0*	70	19
One woman, two or more children	*3*	750	253
One man, one woman	*31*	7,610	2,199
Retired households mainly dependent on state pensions[1]	*3*	700	219
Other retired households	*8*	1,900	580
Non-retired households	*21*	5,020	1,399
Two men or two women	*2*	530	146
Two adults with children	*20*	4,970	1,522
One man one woman, one child	*7*	1,740	507
Two men or two women, one child	*0*	100	29
One man one woman, two children	*9*	2,210	691
Two men or two women, two children	*0*	10	5
One man one woman, three children	*3*	660	211
Two men or two women, three children	*0*	10	4
Two adults, four children	*1*	180	59
Two adults, five children	*0*	30	11
Two adults, six or more children	*0*	20	5
Three adults	*7*	1,700	358
Three adults with children	*3*	610	164
Three adults, one child	*2*	390	98
Three adults, two children	*1*	140	43
Three adults, three children	*0*	50	14
Three adults, four or more children	*0*	30	9
Composition of household (cont)			
Four adults	*2*	430	92
Four adults, one child	*1*	150	37
Four adults, two or more children	*0*	70	20
Five adults	*0*	90	22
Five adults, one or more children	*0*	40	10
All other households without children	*0*	20	5
All other households with children	*0*	10	3
Number of economically active persons in household			
No person	*33*	7,930	2,354
One person	*27*	6,620	1,922
More than one person	*40*	9,790	2,651
Two persons	*31*	7,590	2,140
Three persons	*7*	1,630	384
Four persons	*2*	510	112
Five persons	*0*	60	14
Six or more persons	*0*	10	1
Households with married women	*51*	12,330	3,553
Households with married women economically active	*28*	6,920	1,937
With no dependent children	*15*	3,630	943
With dependent children	*14*	3,300	994
One child	*5*	1,270	361
Two children	*6*	1,560	484
Three children	*2*	380	121
Four or more children	*0*	90	28
Households with married women not economically active	*22*	5,410	1,616
With no dependent children	*17*	4,190	1,232
With dependent children	*5*	1,220	384
One child	*2*	400	118
Two children	*2*	500	160
Three children	*1*	200	65
Four or more children	*0*	120	41
Economic status of household reference person			
Economically active	*62*	15,120	4,229
Employee at work	*52*	12,670	3,537
Full-time	*44*	10,750	2,985
Part-time	*8*	1,920	552
Government-supported training	*0*	40	11
Unemployed	*2*	490	139
Self-employed	*8*	1,920	542
Economically inactive	*38*	9,220	2,696

9.1 Characteristics of households (cont.) 2002-03

based on weighted data

	% * of all house-holds	Grossed number of house-holds (000s)	House-holds in sample (number)
Age of household reference person			
15 and under 20 years	0	50	19
20 and under 25 years	3	770	198
25 and under 30 years	7	1,720	463
30 and under 35 years	10	2,370	687
35 and under 40 years	10	2,500	722
40 and under 45 years	10	2,380	695
45 and under 50 years	9	2,280	648
50 and under 55 years	9	2,200	611
55 and under 60 years	9	2,190	602
60 and under 65 years	7	1,650	469
65 and under 70 years	7	1,680	528
70 and under 75 years	7	1,590	478
75 and under 80 years	5	1,300	381
80 and under 85 years	4	1,050	273
85 and under 90 years	2	460	112
90 years or more	1	160	41
Government Office Regions and Countries 2001-02 -- 2002-03			
United Kingdom	100	24,400	14,400
North East	4	1,020	632
North West	11	2,780	1,599
Yorkshire and the Humber	9	2,130	1,159
East Midlands	7	1,720	974
West Midlands	9	2,120	1,203
East	9	2,230	1,278
London	12	2,850	1,283
South East	14	3,390	1,955
South West	9	2,160	1,282
England	84	20,390	11,365
Wales	5	1,250	712
Scotland	9	2,140	1,207
Northern Ireland	3	620	1,116
Socio-economic class of household reference person[4]			
Higher managerial and professional	11	2,670	707
Large employers/higher managerial	4	1,070	287
Higher professional	7	1,590	420
Lower managerial and professional	17	4,220	1,174
Intermediate	6	1,440	419
Small employers	6	1,380	396
Lower supervisory	7	1,640	464
Semi-routine	7	1,810	512
Routine	7	1,640	472
Long-term unemployed[5]	1	340	91

	% * of all house-holds	Grossed number of house-holds (000s)	House-holds in sample (number)
GB urban/rural areas 2001-02 -- 2002-03			
Urban			
London built-up area	14	3,330	1,564
Other metropolitan built up areas	15	3,600	1,968
Other urban:			
population over 250k	12	2,970	1,707
population 100k to 250k	11	2,560	1,470
population 25k to 100k	14	3,270	1,884
population 10k to 25k	9	2,150	1,242
population 3k to 10k	8	1,900	1,101
Rural	17	4,000	2,348
Tenure of dwelling[2]			
Owners			
Owned outright	29	7,090	2,047
Buying with a mortgage	41	10,030	2,836
All	70	17,130	4,883
Social rented from			
Council	14	3,420	1,012
Registered social landlord	6	1,440	400
All	20	4,850	1,412
Private rented			
Rent free	1	350	100
Rent paid, unfurnished	6	1,440	393
Rent paid, furnished	2	580	139
All	10	2,360	632
Households with durable goods			
Car/van	74	18,070	5,149
One	44	10,610	3,089
Two	25	5,980	1,685
Three or more	6	1,480	375
Central heating, full or partial	93	22,610	6,462
Fridge-freezer or deep freezer	96	23,290	6,618
Washing machine	94	22,980	6,558
Tumble dryer	56	13,660	3,909
Dishwasher	29	7,070	2,044
Microwave oven	87	21,200	6,044
Telephone	94	22,770	6,466
Mobile phone	70	17,120	4,792
Video recorder	90	21,980	6,261
Satellite receiver[3]	45	10,970	3,091
Compact disc player	83	20,270	5,721
Home computer	55	13,430	3,718
Internet connection	45	10,940	3,020

* Based on grossed number of households
1 Mainly dependent on state pension and not economically active - see appendix D.
2 See footnotes in Table 4.10.
3 Includes digital and cable receivers.
4 Excludes economically inactive.
5 Includes those who have never worked and full-time students.

9.2 Characteristics of persons

2002-03

based on weighted data

	Males				Females				All persons		
	Percentage* of		Grossed number of persons (000s)	Persons in the sample (number)	Percentage* of		Grossed number of persons (000s)	Persons in the sample (number)	%* of	Grossed number of persons (000s)	Persons in the sample (number)
	all males	all persons			all females	all persons			all persons		
All persons	100	49	28,203	7,949	100	51	29,787	8,637	100	57,990	16,586
Adults	76	37	21,410	5,828	78	40	23,290	6,622	77	44,700	12,450
Persons aged under 60	57	28	16,154	4,273	56	29	16,688	4,733	57	32,841	9,006
Persons aged 60 or under 65	5	2	1,375	381	5	2	1,442	432	5	2,818	813
Persons aged 65 or under 70	5	2	1,281	400	5	2	1,391	438	5	2,673	838
Persons aged 70 or over	9	4	2,599	774	13	6	3,769	1,019	11	6,368	1,793
Children	24	12	6,793	2,121	22	11	6,498	2,015	23	13,291	4,136
Children under 2 years of age	2	1	701	223	2	1	640	207	2	1,341	430
Children aged 2 or under 5	4	2	1,047	340	3	2	1,008	324	4	2,054	664
Children aged 5 or under 16	15	7	4,196	1,332	14	7	4,096	1,278	14	8,292	2,610
Children aged 16 or under 18	3	1	850	226	3	1	754	206	3	1,604	432
Economic activity											
Persons active (aged 16 or over)	55	27	15,641	4,154	45	23	13,422	3,724	50	29,063	7,878
Persons not active	45	22	12,562	3,795	55	28	16,366	4,913	50	28,928	8,708
Men 65 or over and women 60 or over	13	6	3,611	1,089	20	10	6,023	1,725	17	9,634	2,814
Others (Including children under 16)	32	15	8,951	2,706	35	18	10,343	3,188	33	19,294	5,894

* Based on grossed number of households

9.3 Percentage of households with durable goods, 1970 to 2002-03

	Car/ van	Central heating[1]	Washing machine	Tumble dryer	Dish-washer	Micro-wave	Tele-phone	Mobile phone	Video recorder	Satellite receiver[2]	Cd player	Home computer	Internet connec-tion
1970	52	30	65	--	--	--	35	--	--	--	--	--	--
1975	57	47	72	--	--	--	52	--	--	--	--	--	--
1980	60	59	79	--	--	--	72	--	--	--	--	--	--
1985	63	69	83	--	--	--	81	--	30	--	--	13	--
1990	67	79	86	--	--	--	87	--	61	--	--	17	--
1994-95	69	84	89	50	18	67	91	--	76	--	46	--	--
1995-96	70	85	91	50	20	70	92	--	79	--	51	--	--
1996-97	69	87	91	51	20	75	93	16	82	19	59	27	--
1997-98	70	89	91	51	22	77	94	20	84	26	63	29	--
1998-99	72	89	92	51	24	80	95	26	86	27	68	32	9
1998-99*	72	89	92	51	23	79	95	27	85	28	68	33	10
1999-2000*	71	90	91	52	23	80	95	44	86	32	72	38	19
2000-01*	72	91	92	53	25	84	93	47	87	40	77	44	32
2001-02[3] *	74	92	93	54	27	86	94	64	90	43	80	49	39
2002-03*	74	93	94	56	29	87	94	70	90	45	83	55	45

1 Full or partial.
2 Includes digital and cable receivers
— Data not available.
* Based on weighted data and including children's expenditure

9.4 Percentage of households with durable goods by income group and household composition

2002-03

based on weighted data

	Central heating[2]	Washing machine	Tumble dryer	Micro-wave	Dish-washer	CD player
All households	93	94	56	87	29	83
Gross income decile group						
Lowest ten per cent	87	80	33	76	8	58
Second decile group	89	87	38	81	9	60
Third decile group	90	92	41	84	14	71
Fourth decile group	92	94	54	86	18	82
Fifth decile group	91	97	57	89	23	86
Sixth decile group	95	97	64	90	28	92
Seventh decile group	94	99	62	90	34	94
Eighth decile group	96	100	66	91	42	95
Ninth decile group	97	99	69	92	48	98
Highest ten per cent	97	99	76	92	67	97
Household composition						
One adult, retired households[3]	88	74	29	71	5	35
One adult, non-retired households	89	88	40	82	15	84
One adult, one child	93	97	46	88	10	93
One adult, two or more children	94	97	61	89	20	95
One man and one woman, retired households[3]	87	93	37	81	11	55
One man and one woman, non-retired households	94	98	61	89	34	92
One man and one woman, one child	95	100	73	93	39	98
One man and one woman, two or more children	96	99	76	93	47	96
All other households without children	93	96	54	90	37	92
All other households with children	90	97	62	94	36	89

	Home computer	Internet connection	Tele-phone	Mobile phone	Satellite receiver[4]	Video recorder
All households	55	45	94	70	45	90
Gross income decile group						
Lowest ten per cent	19	12	83	37	23	73
Second decile group	21	14	89	45	26	80
Third decile group	31	22	93	53	33	85
Fourth decile group	43	30	92	68	44	91
Fifth decile group	53	41	94	77	44	93
Sixth decile group	60	48	94	80	51	95
Seventh decile group	70	56	96	81	52	96
Eighth decile group	81	68	98	86	58	98
Ninth decile group	84	74	98	88	62	97
Highest ten per cent	91	85	99	88	59	95
Household composition						
One adult, retired households[3]	5	4	96	14	15	63
One adult, non-retired households	45	36	85	68	31	86
One adult, one child	58	34	82	77	45	93
One adult, two or more children	56	36	78	84	50	96
One man and one woman, retired households[3]	13	8	99	37	25	85
One man and one woman, non-retired households	63	55	95	82	52	95
One man and one woman, one child	79	68	97	86	62	97
One man and one woman, two or more children	83	69	95	86	63	98
All other households without children	69	57	93	84	48	94
All other households with children	70	52	91	82	53	97

1 See table 9.5 for number of recording households.
2 Full or partial.
3 Mainly dependent on state pension and not economically active - see appendix D.
4 Includes digital and cable receivers

9.5 Percentage of households with cars by income group, tenure and household composition 2002-03

based on weighted data

	One car/van	Two cars/vans	Three or more cars/vans	All with cars/vans	Grossed number of households (000s)	Households in the sample (number)
All households	*44*	*25*	*6*	*74*	24,350	6,927
Gross income decile group						
Lowest ten per cent	*24*	*1*	*1*	*26*	2,440	701
Second decile group	*37*	*3*	*1*	*41*	2,430	724
Third decile group	*49*	*7*	*1*	*57*	2,440	732
Fourth decile group	*60*	*11*	*2*	*73*	2,430	715
Fifth decile group	*60*	*19*	*2*	*81*	2,440	719
Sixth decile group	*57*	*26*	*5*	*88*	2,440	697
Seventh decile group	*51*	*37*	*4*	*92*	2,430	687
Eighth decile group	*44*	*44*	*7*	*95*	2,440	672
Ninth decile group	*29*	*51*	*16*	*95*	2,430	655
Highest ten per cent	*25*	*48*	*22*	*95*	2,430	625
Tenure of dwelling[1]						
Owners						
Owned outright	*50*	*22*	*7*	*79*	7,090	2,047
Buying with a mortgage	*45*	*38*	*9*	*92*	10,030	2,836
All	*47*	*31*	*8*	*86*	17,130	4,883
Social rented from						
Council	*32*	*5*	*2*	*39*	3,420	1,012
Registered social landlord[2]	*30*	*6*	*1*	*37*	1,440	400
All	*31*	*6*	*1*	*38*	4,850	1,412
Private rented						
Rent free	*47*	*15*	*0*	*63*	350	100
Rent paid, unfurnished	*49*	*15*	*2*	*67*	1,440	393
Rent paid, furnished	*31*	*12*	*1*	*44*	580	139
All	*45*	*14*	*2*	*61*	2,360	632
Household composition						
One adult, retired mainly dependent on state pensions[3]	*16*	*0*	*0*	*16*	1,640	459
One adult, other retired	*44*	*1*	*0*	*45*	1,470	416
One adult, non-retired	*58*	*6*	*1*	*64*	3,490	978
One adult, one child	*47*	*3*	*0*	*50*	700	224
One adult, two or more children	*46*	*4*	*0*	*51*	820	272
One man and one woman, retired dependent on state pensions[3]	*56*	*4*	*0*	*60*	700	219
One man and one woman, other retired	*65*	*19*	*2*	*85*	1,900	580
One man and one woman, non-retired	*44*	*39*	*7*	*91*	5,020	1,399
One man and one woman, one child	*44*	*39*	*8*	*91*	1,740	507
One man and one woman, two children	*42*	*47*	*5*	*94*	2,210	691
One man and one woman, three children	*41*	*44*	*3*	*88*	660	211
Two adults, four or more children	*38*	*32*	*9*	*80*	230	75
Three adults	*26*	*39*	*22*	*87*	1,700	358
Three adults, one or more children	*33*	*38*	*18*	*89*	610	164
All other households without children	*30*	*29*	*18*	*77*	1,080	265
All other households with children	*32*	*21*	*23*	*77*	390	108

1 See footnotes in Table 4.10.
2 Formerly housing association
3 Mainly dependent on state pension and not economically active - see appendix D.

9.6 Percentage of households with durable goods by UK Countries and Government Office Regions

2001-02 – 2002-03

based on weighted data

	North East	North West	Yorks and the Humber	East Midlands	West Midlands	East	London
Average number of grossed households (thousands)	**1,020**	**2,780**	**2,130**	**1,720**	**2,120**	**2,230**	**2,850**
Total number of households in sample (over 2 years)	**632**	**1,599**	**1,159**	**974**	**1,203**	**1,278**	**1,283**
Percentage of households by Government Office Region and country							
Car/van	62	70	70	79	75	82	65
One	38	42	45	44	43	46	43
Two	19	23	20	27	26	28	18
Three or more	5	5	5	8	7	7	5
Central heating full or partial	98	91	88	94	90	94	94
Fridge-freezer or deep freezer	94	92	94	97	94	95	92
Washing machine	97	95	95	97	96	96	95
Tumble dryer	55	56	54	59	58	55	44
Dishwasher	20	23	22	28	24	34	28
Microwave	90	88	90	89	89	85	80
Telephone	95	93	93	96	93	96	94
Mobile phone	67	64	72	67	70	67	65
Video recorder	92	91	91	92	91	91	87
Satellite receiver[2]	47	46	42	46	44	43	43
CD player	82	80	78	83	80	84	81
Home computer	45	49	47	53	52	58	59
Internet connection	36	40	37	44	38	48	49

	South East	South West	England	Wales	Scotland	Northern Ireland	United Kingdom
Average number of grossed households (thousands)	**3,390**	**2,160**	**20,390**	**1,250**	**2,140**	**620**	**24,400**
Total number of households in sample (over 2 years)	**1,955**	**1,282**	**11,365**	**712**	**1,207**	**1,116**	**14,400**
Percentage of households by Government Office Region and country							
Car/van	83	82	75	74	66	71	74
One	45	46	44	46	44	47	44
Two	29	28	25	23	19	21	24
Three or more	9	8	6	5	4	3	6
Central heating full or partial	94	91	92	91	92	97	92
Fridge-freezer or deep freezer	94	93	94	94	96	96	94
Washing machine	97	96	96	96	94	91	95
Tumble dryer	58	55	55	57	56	52	55
Dishwasher	39	30	29	22	24	33	28
Microwave	86	83	86	89	88	86	86
Telephone	95	94	94	90	93	91	94
Mobile phone	75	64	68	63	67	50	67
Video recorder	91	88	90	90	90	87	90
Satellite receiver[1]	42	38	43	54	45	40	44
CD player	85	80	82	81	85	71	82
Home computer	58	50	53	46	47	41	52
Internet connection	50	39	43	34	39	33	42

1 Includes digital and cable receivers

9.7 Percentage of households by size, composition and age in each income decile group 2002-03

based on weighted data

	Lowest ten per cent	Second decile group	Third decile group	Fourth decile group	Fifth decile group	Sixth decile group
Lower boundary of group (£ per week)		123	188	259	341	435
Grossed number of households (thousands)	**2,440**	**2,430**	**2,440**	**2,430**	**2,440**	**2,440**
Number of households in the sample	**701**	**724**	**732**	**715**	**719**	**697**
Size of household						
One person	78	53	37	32	23	19
Two persons	17	29	47	39	44	42
Three persons	4	11	8	15	18	17
Four persons	1	6	4	7	10	14
Five persons	0	1	3	3	4	7
Six or more persons	0	0	1	3	3	2
All sizes	100	100	100	100	100	100
Household composition						
One adult, retired mainly dependent on state pensions[1]	45	18	4	0	0	0
One adult, other retired	3	23	16	9	4	2
One adult, non-retired	30	13	16	23	18	17
One adult, one child	11	4	4	3	3	2
One adult, two or more children	2	11	5	5	4	4
One man and one woman, retired mainly dependent on state pensions[1]	0	15	11	2	0	0
One man and one woman, other retired	0	2	18	16	17	9
One man and one woman, non-retired	6	8	11	14	21	28
One man and one woman, one child	1	2	2	7	8	9
One man and one woman, two children	1	2	2	5	8	11
One man and one woman, three children	0	0	2	3	3	5
Two adults, four or more children	0	0	0	2	2	1
Three adults	2	1	2	4	7	5
Three adults, one or more children	0	0	1	1	1	3
All other households without children	0	2	3	4	3	3
All other households with children	0	0	1	2	1	1
All compositions	100	100	100	100	100	100
Age of household reference person						
15 and under 20 years	1	0	1	1	0	0
20 and under 25 years	6	5	2	4	4	3
25 and under 30 years	5	4	5	7	8	10
30 and under 35 years	5	5	4	8	11	14
35 and under 40 years	6	5	5	9	10	13
40 and under 45 years	5	4	6	9	8	9
45 and under 50 years	5	4	4	8	9	11
50 and under 55 years	6	4	6	6	7	9
55 and under 60 years	8	4	7	9	10	9
60 and under 65 years	7	8	8	7	9	8
65 and under 70 years	8	11	13	11	8	5
70 and under 75 years	10	14	14	7	7	4
75 and under 80 years	10	13	10	8	4	2
80 and under 85 years	10	11	10	4	4	1
85 and under 90 years	6	6	4	1	1	1
90 years or more	2	2	1	1	0	0
All ages	100	100	100	100	100	100

1 Mainly dependent on state pension and not economically active - see appendix D.

9.7 Percentage of households by size, composition and age in each income decile group (cont.)

2002-03

based on weighted data

	Seventh decile group	Eighth decile group	Ninth decile group	Highest ten per cent	All house-holds
Lower boundary of group (£ per week)	541	662	821	1085	
Grossed number of households (thousands)	**2,430**	**2,440**	**2,430**	**2,430**	**24,350**
Number of households in the sample	**687**	**672**	**655**	**625**	**6,927**
Size of household					
One person	11	9	5	4	27
Two persons	39	40	38	29	36
Three persons	20	26	24	27	17
Four persons	21	18	25	28	13
Five persons	6	6	7	9	4
Six or more persons	3	2	2	4	2
All sizes	100	100	100	100	100
Household composition					
One adult, retired mainly dependent on state pensions[1]	0	0	0	0	7
One adult, other retired	2	1	0	0	6
One adult, non-retired	10	8	4	4	14
One adult, one child	1	1	0	0	3
One adult, two or more children	1	1	0	1	3
One man and one woman, retired mainly dependent on state pensions[1]	0	0	0	0	3
One man and one woman, other retired	6	4	4	2	8
One man and one woman, non-retired	28	33	32	25	21
One man and one woman, one child	11	11	9	12	7
One man and one woman, two children	16	14	17	15	9
One man and one woman, three children	4	4	3	4	3
Two adults, four or more children	1	1	1	2	1
Three adults	8	13	14	14	7
Three adults, one or more children	3	5	6	5	3
All other households without children	7	3	7	12	4
All other households with children	2	2	2	4	2
All compositions	100	100	100	100	100
Age of household reference person					
15 and under 20 years	0	0	0	0	0
20 and under 25 years	3	2	2	2	3
25 and under 30 years	8	9	8	7	7
30 and under 35 years	14	13	12	11	10
35 and under 40 years	13	13	16	12	10
40 and under 45 years	11	17	15	14	10
45 and under 50 years	11	9	13	19	9
50 and under 55 years	12	13	12	16	9
55 and under 60 years	10	11	11	11	9
60 and under 65 years	7	5	5	4	7
65 and under 70 years	4	4	3	1	7
70 and under 75 years	3	3	2	1	7
75 and under 80 years	3	2	0	1	5
80 and under 85 years	2	0	0	0	4
85 and under 90 years	0	0	0	0	2
90 years or more	0	0	0	0	1
All ages	100	100	100	100	100

1 Mainly dependent on state pension and not economically active - see appendix D.

9.8 Percentage of households by economic activity, tenure and socio-economic class in each income decile group — 2002-03

based on weighted data

	Lowest ten per cent	Second decile group	Third decile group	Fourth decile group	Fifth decile group	Sixth decile group
Lower boundary of group (£ per week)		123	188	259	341	435
Grossed number of households (thousands)	**2,440**	**2,430**	**2,440**	**2,430**	**2,440**	**2,440**
Number of households in the sample	**701**	**724**	**732**	**715**	**719**	**697**
Number of economically active persons in household						
No person	*81*	*80*	*68*	*39*	*25*	*12*
One person	*17*	*17*	*25*	*46*	*47*	*40*
Two persons	*2*	*3*	*7*	*13*	*24*	*43*
Three persons	*0*	*0*	*1*	*2*	*3*	*4*
Four or more persons	*0*	*0*	*0*	*0*	*1*	*1*
All economically active persons	*100*	*100*	*100*	*100*	*100*	*100*
Tenure of dwelling[1]						
Owners						
Owned outright	*26*	*37*	*45*	*34*	*35*	*27*
Buying with a mortgage	*9*	*8*	*13*	*29*	*37*	*52*
All	*35*	*44*	*58*	*64*	*72*	*78*
Social rented from						
Council	*34*	*30*	*22*	*19*	*13*	*9*
Registered social landlord [2]	*16*	*14*	*9*	*8*	*5*	*2*
All	*50*	*45*	*31*	*27*	*18*	*11*
Private rented						
Rent free	*3*	*2*	*3*	*2*	*1*	*1*
Rent paid, unfurnished	*8*	*7*	*6*	*5*	*6*	*7*
Rent paid, furnished	*3*	*2*	*2*	*2*	*2*	*2*
All	*15*	*11*	*11*	*9*	*10*	*10*
All tenures	*100*	*100*	*100*	*100*	*100*	*100*
Socio-economic class						
Higher managerial and professional						
Large employers/higher managerial	*0*	*0*	*0*	*1*	*1*	*2*
Higher professional	*0*	*0*	*1*	*2*	*3*	*5*
Lower managerial and professional	*2*	*1*	*3*	*8*	*13*	*21*
Intermediate	*1*	*0*	*5*	*9*	*9*	*9*
Small employers	*2*	*4*	*4*	*6*	*8*	*8*
Lower supervisory	*1*	*1*	*3*	*5*	*8*	*12*
Semi-routine	*3*	*4*	*8*	*11*	*10*	*13*
Routine	*4*	*3*	*5*	*10*	*10*	*12*
Long-term unemployed [4]	*5*	*3*	*0*	*2*	*1*	*1*
Other[3]	*81*	*83*	*71*	*48*	*36*	*18*
All occupational groups	*100*	*100*	*100*	*100*	*100*	*100*

1 See footnotes in Table 4.10.
2 Formerly housing association
3 See Appendix D for definition of NS-SEC
4 Includes those who have never worked and full-time students.

9.8 Percentage of households by economic activity, tenure and socio-economic class in each income decile group (cont.) 2002-03

based on weighted data

	Seventh decile group	Eighth decile group	Ninth decile group	Highest ten per cent	All house-holds
Lower boundary of group (£ per week)	541	662	821	1085	
Grossed number of households (thousands)	**2,430**	**2,440**	**2,430**	**2,430**	**24,350**
Number of households in the sample	**687**	**672**	**655**	**625**	**6,927**
Number of economically active persons in household					
No person	*9*	*5*	*5*	*2*	*33*
One person	*30*	*21*	*13*	*15*	*27*
Two persons	*52*	*58*	*59*	*51*	*31*
Three persons	*8*	*14*	*17*	*19*	*7*
Four or more persons	*1*	*2*	*6*	*13*	*2*
All economically active persons	*100*	*100*	*100*	*100*	*100*
Tenure of dwelling[1]					
Owners					
Owned outright	*26*	*21*	*20*	*20*	*29*
Buying with a mortgage	*58*	*67*	*68*	*71*	*41*
All	*84*	*88*	*88*	*91*	*70*
Social rented from					
Council	*5*	*3*	*2*	*2*	*14*
Registered social landlord [2]	*2*	*2*	*1*	*0*	*6*
All	*7*	*5*	*3*	*2*	*20*
Private rented					
Rent free	*1*	*0*	*1*	*0*	*1*
Rent paid, unfurnished	*5*	*5*	*5*	*3*	*6*
Rent paid, furnished	*3*	*2*	*3*	*3*	*2*
All	*9*	*7*	*8*	*7*	*10*
All tenures	*100*	*100*	*100*	*100*	*100*
Socio-economic class					
Higher managerial and professional					
Large employers/higher managerial	*4*	*5*	*10*	*21*	*4*
Higher professional	*8*	*11*	*13*	*23*	*7*
Lower managerial and professional	*26*	*31*	*36*	*32*	*17*
Intermediate	*8*	*7*	*6*	*5*	*6*
Small employers	*6*	*7*	*8*	*4*	*6*
Lower supervisory	*12*	*12*	*9*	*4*	*7*
Semi-routine	*10*	*8*	*4*	*2*	*7*
Routine	*10*	*7*	*4*	*3*	*7*
Long-term unemployed[4]	*1*	*1*	*0*	*0*	*1*
Other[3]	*16*	*11*	*9*	*5*	*38*
All occupational groups	*100*	*100*	*100*	*100*	*100*

1 See footnotes in Table 4.10.
2 Formerly housing association
3 See Appendix D for definition of NS-SEC
4 Includes those who have never worked and full-time students.

Appendices

Appendix A	Description and response rate of the survey
Appendix B	Uses of the survey
Appendix C	Standard errors and estimates of precision
Appendix D	Definitions
Appendix E	Changes in definition, 1991 to 2002-03
Appendix F	Differential grossing
Appendix G	Index to tables in reports on the FES/EFS in 1994-95 to 2002-03

Appendix A
Description and response rate of the survey

The survey

The Expenditure and Food Survey (EFS) is a voluntary sample survey of private households. The basic unit of the survey is the household. In the 2001-02 survey the EFS adopted the harmonised definition used in other government household surveys: a group of people living at the same address with common housekeeping, that is sharing household expenses such as food and bills, or sharing a living room (see Appendix D). The previous definition differed from the harmonised definition by requiring both common housekeeping **and** a shared living room. This resulted in the EFS having slightly more one person households and fewer large households than the other surveys.

Each individual aged 16 or over in the household visited is asked to keep diary records of daily expenditure for two weeks. Information about regular expenditure, such as rent and mortgage payments, is obtained from a household interview along with retrospective information on certain large, infrequent expenditures such as those on vehicles. Since 1998-99 the results have also included information from simplified diaries kept by children aged between 7 and 15. The effects were shown in Appendix F of Family Spending for 1998-99 and again for 1999-2000. The analysis is not repeated this year as inclusion of the data is now a standard feature of the survey.

Detailed questions are asked about the income of each adult member of the household. In addition, personal information such as age, sex and marital status is recorded for each household member. Paper versions of the computerised household and income questionnaires can be obtained from the address given in the Introduction.

The survey has been conducted each year since 1957. The survey is continuous, interviews being spread evenly over the year to ensure that seasonal effects are covered. From time to time changes are made to the information sought. Some changes reflect new forms of expenditure or new sources of income, especially benefits. Others are the result of new requirements by the survey's users. An important example is the re-definition of housing costs for owner occupiers in 1992 (see Appendix E).

The sample design

The EFS sample for Great Britain is a multi-stage stratified random sample with clustering. It is drawn from the Small Users file of the Postcode Address File - the Post Office's list of addresses. All Scottish offshore islands and the Isles of Scilly are excluded from the sample because of excessive interview travel costs. Postal sectors (ward size) are the primary sample unit. Six hundred and seventy two postal sectors are randomly selected during the year after being arranged in strata defined by Government Office Regions (sub-divided into metropolitan and non-metropolitan areas) and two 1991 Census variables - socio-economic group and ownership of cars. These were new stratifiers introduced for the 1996-97 survey. The Northern Ireland sample is drawn as a random sample of addresses from the Valuation and Lands Agency list.

Response to the survey

Great Britain

Some 12,096 households are selected each year for the EFS in Great Britain, but it is never possible to get full response. A small number cannot be contacted at all, and in other households one or more members decline to co-operate. Six thousand two hundred and fifteen households in Great Britain co-operated fully in the survey in 2002-03, that is they answered the household questionnaire and all adults in the household answered the income questionnaire and kept the expenditure diary. A further 127 households provided sufficient information to be included as valid responses. The overall response rate for the 2002-03 EFS was 58 per cent in Great Britain. This compares with 62 per cent in 2001-02.

Details of response are shown in the following table.

Response in 2002-03 - Great Britain

		No of households or addresses	Percentage of effective sample
i.	Sampled addresses	12,096	-
ii.	Ineligible addresses: businesses, institutions, empty, demolished/derelict	1,225	-
iii.	Extra households (multi-household addresses)	148	-
iv.	Total eligible (i.e. i less ii, plus iii)	11,019	100.0
v.	Co-operating households (which icludes 127 partials)	6,342	57.5
vi.	Refusals	3,858	35.0
vii.	Households at which no contact could be obtained	819	7.4

Northern Ireland

In the Northern Ireland survey, the eligible sample was 1,039 households. The number of co-operating households who provided usable data was 585, giving a response rate of 56 per cent. Northern Ireland is over-sampled in order to provide a large enough sample for some separate analysis. The re-weighting procedure compensates for the over-sampling.

The fieldwork

The fieldwork is conducted by the Office for National Statistics (ONS) in Great Britain and by the Northern Ireland Statistics and Research Agency of the Department of Finance and Personnel in Northern Ireland using almost identical questionnaires. Households at the selected addresses are visited and asked to co-operate in the survey. In order to maximise response, interviewers make at least four separate calls, and sometimes many more, at different times of day on households which are difficult to contact. Interviews are conducted by Computer Assisted Personal Interviewing (CAPI) using portable computers. During the interview information is collected about the household, about certain regular payments such as rent, gas, electricity and telephone accounts, about expenditure on certain large items (for example vehicle purchases over the previous 12 months), and about income. Each individual aged 16 or over in the household keeps a detailed record of expenditure every day for two weeks. Children aged between 7 and 15 are also asked to keep a simplified diary of daily expenditure (though not in the Northern Ireland enhanced sample). In 2002-03 a total of 1954 children aged between 7 and 15 in responding households in Great Britain were asked to complete expenditure diaries; only 40 or about 2 per cent did not do so. This number includes both refusals and children who had no expenditure during the two weeks. Information provided by all members of the household is kept strictly confidential. If all persons aged 16 and over in the household co-operate each is subsequently paid £10 for the trouble involved. Children who keep a diary are given a £5 payment.

In the last two months of the 1998-99 survey, as an experiment, a small book of postage stamps was enclosed with the introductory letter sent to every address. It seemed to help with response and the measure has become a permanent feature of the survey. It is difficult to quantify the exact effect on response but the cognitive work that was carried out as part of the Expenditure and Food Survey development indicated that it was having a positive effect.

A new strategy for reissues adopted in 1999-2000 was continued in 2002-03. Addresses where there had been no contact or a refusal, but were judged suitable for reissue, were accumulated to form complete batches consisting only of reissues. The interviewers dealing with them were specially selected and given extra briefing. The information from households converted from non-responding to responding was included with the data for the quarter of the year when the interview was carried out. The increase in response rate, however, was attributed to the original month of issue. Some 920 addresses were reissued, of which 148 were converted into fully responding households, which added 1.3 percentage points to the response rate.

Eligible response

Under EFS rules, a refusal by just one person to respond to the income section of the questionnaire invalidates the response of the whole household response. Similarly, a refusal by the household's main shopper to complete the two-week expenditure diary also results in an invalid response.

Proxy Interviews – While questions about general household affairs are put to all household members or to a main household informant, questions about work and income are put to the individual members of the household. Where a member of the household is not present during the household interview, another member of the household (e.g. spouse) may be able to provide information about the absent person. The individual's interview is then identified as a proxy interview.

In 2001-02, the EFS began including households that contained a proxy interview. In that year, 12 per cent of all responding households contained at least one proxy interview. In 2002-3, 14 per cent of all responding households did so. Analysis of the 2001-02 data has revealed that the inclusion of proxy interviews increased response from above average income households. For the 2001-02 survey, the average gross normal weekly household income was some three per cent higher than it would have been if proxy interviews had not been accepted. The analysis showed a similar difference for average total expenditure.

Short Income – All adult members of a household must supply information about their income. This information is quite detailed, and a small number of respondents are reluctant to provide it. For these people, we accept responses to a reduced set of questions, called short income.

From 2001-02, the EFS began including households that contained a short income section. In that year, 0.5 per cent of households contained at least one short income response. In 2002-3, 0.3 per cent did so.

Reliability

Great care is taken in collecting information from households and comprehensive checks are applied during processing, so that errors in recording and processing are minimised. The main factors that affect the reliability of the survey results are sampling variability, non-response bias and some incorrect reporting of certain items of expenditure and income. Measures of sampling variability are given alongside some results in this report and are discussed in detail in Appendix C.

The households which decline to respond to the survey tend to differ in some respects from those which co-operate. It is therefore possible that their patterns of expenditure and income also differ. A comparison was made of the households responding in the 1991 FES with those not responding, based on information from the 1991 Census of Population (A comparison of the Census characteristics of respondents and non-respondents to the 1991 FES by K Foster, ONS Survey Methodology Bulletin No. 38, Jan 1996). Results from the study indicate that response was lower than average in Greater London, higher in non-metropolitan areas and that non-response tended to increase with increasing age of the head of the household, up to age 65. Households that contained three or more adults, or where the head was born outside the United Kingdom or was classified to an ethnic minority group, were also more likely than others to be non-responding. Non-response was also above average where the head of the household had no post-school qualifications, was self-employed, or was in a manual social class group. The data are now re-weighted to compensate for the main non-response biases identified from the 1991 Census comparison, as described in Appendix F. ONS is currently undertaking a similar comparative exercise, with the 2001 Census data, which will result in an update of the non-response weights.

Checks are included in the CAPI program which are applied to the responses given during the interview. Other procedures are also in place to ensure that users are provided with high quality data. For example, quality control is carried out to ensure that any outliers are genuine, and checks are made on any unusual changes in average spending compared with the previous year.

When aspects of the survey change, rigorous tests are used to ensure the proposed changes are sensible and work both in the field and on the processing system. For example, in 1996-97 an improved set of questions was introduced on income from self-employment. This was developed by focus groups and then tested by piloting before being introduced into the main survey.

The information obtained by the survey does not permit the construction of household accounts in the form of an income-expenditure balance sheet. The definitions of weekly household expenditure and income used are such that it is not to be expected that expenditure and income will balance, either for an individual household or even when averaged over a group of households. Hence, the difference between expenditure and income is not a measure of savings or dis-savings.

Imputation of missing information

Although EFS response is generally based on complete households responding, there are areas in the survey for which missing information can be imputed. This falls into two broad categories:

(i) Specific items of information missing from a response. These missing values are imputed on a case by case basis using other information collected in the interview. The procedure is used, for example, for council tax payments and for interest received on savings.

(ii) Imputation of a complete diary case. Where a response is missing a diary from a household member, this information is imputed using information from respondents with similar characteristics.

Appendix B
Uses of the survey

EFS Expenditure Data

Retail Prices Index - The main reason, historically, for instituting a regular survey on expenditure by households has been to provide information on spending patterns for the Retail Prices Index (RPI). The RPI plays a vital role in the uprating of state pensions and welfare benefits and in general economic policy and analysis. The RPI measures the change in the cost of a selection of goods and services representative of the expenditure of the vast majority of households. The pattern of expenditure gradually changes from one year to the next, and the composition of the basket needs to be kept up-to-date. Accordingly, regular information is required on spending patterns and much of this is supplied by the EFS. The expenditure weights for the general RPI need to relate to people within given income limits, for which the EFS is the only source of information.

Consumers' expenditure and GDP - EFS data on spending are an important source used in compiling national estimates of consumers' expenditure which are published regularly in United Kingdom National Accounts (ONS Blue Book). Consumers' expenditure estimates feed into the National Accounts and estimates of GDP. They will also provide the weights for the Harmonised Index of Consumer prices (HICP) which is calculated by each member of the European Union, and for Purchasing Power Parities (PPPs) for international price comparisons. EFS data are also used in the estimation of taxes on expenditure, in particular VAT.

Regional accounts - EFS expenditure information is one of the sources used by ONS to derive regional estimates of consumers' expenditure. It is also used in compiling some of the other estimates for the regional accounts.

Pay Review Bodies governing the salaries of HM Armed Forces and the medical and dental professions receive estimates of the minimum valuation of the benefit of a company car. This is based on EFS data on the cost of buying and running private cars.

The Statistical Office of the European Communities (Eurostat) collates information from family budget surveys conducted by the member states. The EFS is the UK's contribution to this. The UK is one of only a few countries with such a regular, continuous and detailed survey.

Other Government uses - The Department of Trade and Industry and the Department for Transport, both use EFS expenditure data in their own fields, e.g. - relating to energy, housing, cars and transport. Several other government publications include EFS expenditure data, such as *Social Trends, Regional Trends* and the *Social Focus* series.

Non-Government uses - There are also numerous users outside Central Government, including academic researchers and business and market researchers.

EFS Income Data

Redistribution of income - EFS information on income and expenditure is used to study how Government taxes and benefits affect household income. The Government's interdepartmental tax benefit model is based on the EFS and enables the economic effects of policy measures to be analysed across households. This model is used by HM Treasury, Inland Revenue and HM Customs and Excise to estimate the impact on different households of possible changes in taxes and benefits.

Non-Government users - As with the expenditure data, EFS income data are also studied extensively outside Government. In particular, academic researchers in the economic and social science areas of many universities use the EFS. For example the Institute for Fiscal Studies uses EFS data in research it carries out both for Government and on its own account to inform public debate.

Other EFS Data

The Department for Environment, Food and Rural Affairs (DEFRA) publishes separate reports using EFS data on food expenditure and consumption, and nutrient intake.

The Office for National Statistics uses the information on access to the Internet in a quarterly analysis of Internet access. The Department for Transport uses EFS data to monitor and forecast levels of car ownership and use, and in studies on the effects of motoring taxes.

Note: Great care is taken to ensure complete confidentiality of information and to protect the identity of EFS households. Only anonymised data are supplied to users.

Appendix C
Standard errors and estimates of precision

Because the EFS is a survey of a sample of households and not of the whole population, the results are liable to differ to some degree from those that would have been obtained if every single household had been covered. Some of the differences will be systematic, in that lower proportions of certain types of household respond than of others. That aspect is discussed in Appendices A and F. This Appendix discusses the effect of sampling variability, the differences in expenditure and income between the households in the sample and in the whole population that arise from random chance. The degree of possible error will depend on the sample size and how widely particular categories of expenditure (or income) vary between households. This "sampling error" is smallest for the average expenditure of large groups of households on items purchased frequently and when the level of spending does not vary greatly between households. Conversely it is largest for small groups of households, and for items purchased infrequently or for which expenditure varies considerably between households. A numerical measure of the likely magnitude of such differences (between the sample estimate and the value of the entire population) is provided by the quantity known as the standard error.

The calculation of standard errors takes into account the fact that the EFS sample is drawn in two stages, first a sample of areas (primary sampling units) then a sample of addresses within each of these areas. The main features of the sample design are described in Appendix A to this Report. The calculation also takes account of the effect of weighting. The two-stage sample increases sampling variability slightly, but the weighting reduces it for some items.

Standard errors for detailed expenditure items are presented in relative terms in **table 7.1** (standard error as a percentage of the average to which it refers). As the calculation of full standard errors is complex, this is the only table where they are shown. **Tables C1** and **C2** in this Appendix show the design factor (DEFT), a measure of the efficiency of the survey's sample design. The DEFT is calculated by dividing the 'full' standard error by the standard error that would have applied if the survey had used a simple random sample ('simple method'). Please note that the DEFTs published in this edition of the report have been revised following a review of the methods applied in previous editions.

Table C1
Percentage standard errors of expenditure of households and number of recording households 2002-03

Commodity or service	Weighted average weekly household expenditure (£)	Percentage standard error: Simple method	Design factor (DEFT)	Percentage standard error: Full method	Households recording expenditure: Recording households in sample	Households recording expenditure: Percentage of all households
All expenditure groups	**348.30**	***1.0***	**1.0**	***1.0***	**6,927**	***100***
Food and non-alcoholic drinks	42.70	*0.7*	1.0	*0.8*	6,898	*100*
Alcoholic drink and tobacco	11.40	*1.8*	1.1	*1.9*	4,529	*65*
Clothing and footwear	22.30	*2.2*	1.0	*2.1*	4,808	*69*
Housing, fuel and power	36.90	*1.6*	1.1	*1.7*	6,891	*99*
Household goods and services	30.20	*2.6*	1.0	*2.6*	6,463	*93*
Health	4.80	*5.1*	1.0	*4.8*	3,554	*51*
Transport	59.20	*1.8*	1.1	*2.0*	6,066	*88*
Communication	10.60	*1.3*	0.9	*1.2*	6,652	*96*
Recreation and culture	56.40	*1.9*	1.1	*2.1*	6,872	*99*
Education	5.20	*7.2*	1.0	*7.5*	729	*11*
Restaurants and hotels	35.40	*1.7*	1.1	*1.8*	6,216	*90*
Miscellaneous	33.10	*1.7*	1.1	*1.8*	6,782	*98*

Please note that the DEFTs published in this edition of the report have been revised following a review of the methods applied in previous editions.

Table C2
Percentage standard errors of income of households and number of recording households 2002-03

Source of income	Weighted average weekly household income (£)	Percentage standard error: Simple method	Design factor (DEFT)	Percentage standard error: Full method	Households recording income: Recording households in sample	Households recording income: Percentage of all households
Gross household income	**552**	***1.7***	**0.8**	***1.3***	**6,927**	***100***
Wages and salaries	374	*1.9*	0.9	*1.6*	4,117	*59*
Self-employment	45	*5.7*	1.0	*5.6*	768	*11*
Investments	19	*16.8*	0.3	*5.0*	4,057	*59*
Annuities and pensions (other than social security benefits)	40	*3.4*	0.9	*3.1*	1,887	*27*
Social security benefits	68	*1.3*	0.8	*1.0*	4,939	*71*
Other sources	7	*6.1*	1.0	*6.3*	1,285	*19*

Please note that the DEFTs published in this edition of the report have been revised following a review of the methods applied in previous editions.

Using the standard errors – confidence intervals

A good way of using standard errors is to calculate 95% confidence intervals from them. Simplifying a little, these can be taken to mean that there is only a 5% chance that the true population value lies outside that confidence interval. The 95% confidence interval is calculated as 1.96 times the standard error on either side of the mean. For example the average expenditure on alcoholic drink and tobacco is £11.40 and the corresponding percentage standard error (full method) is 1.9%. The amount either side of the mean for 95% confidence is then:

1.96 x (1.9 ÷100) x £11.40 = £0.40 (rounded to nearest 10p)
Lower limit is 11.40 – 0.40 = £11.00
Upper limit is 11.40 + 0.40 = £11.80

Similar calculations can be carried out for other estimates of expenditure and income. The 95% confidence intervals for main expenditure categories are given in **Table C3**.

Table C3
95 per cent confidence intervals for average household expenditure 2002-03

Commodity or service	Weighted average weekly household expenditure (£)	95% confidence interval: Lower limit	95% confidence interval: Upper limit
All expenditure groups	**348.30**	**342.10**	**354.40**
Food and non-alcoholic drinks	42.70	42.10	43.40
Alcoholic drink and tobacco	11.40	11.00	11.80
Clothing and footwear	22.30	21.50	23.10
Housing, fuel and power	36.90	35.70	38.10
Household goods and services	30.20	28.70	31.80
Health	4.80	3.80	5.80
Transport	59.20	57.20	61.30
Communication	10.60	10.20	11.00
Recreation and culture	56.40	54.60	58.20
Education	5.20	4.30	6.00
Restaurants and hotels	35.40	34.50	36.30
Miscellaneous	33.10	32.00	34.20

Calculation of standard errors

Simple method

This formula treats the EFS sample as though it had arisen from a much simpler design with no multi-stage sampling, stratification, nor differential sampling and no nonresponse weights. The weights are used but only to estimate the true population standard deviation in what is, in fact, a weighted design. The method of calculation is as follows: Let n be the total number of responding households in the survey, x_r the expenditure on a particular item of the r-th household, w_r be the weight attached to household r, and $\bar{x}$ the average expenditure per household on that item (averaged over the n households). Then the standard error $\bar{x}$, $sesrs$ is given by:

$$sesrs = \sqrt{\frac{\sum_{r=1}^{n} w_r (x_r - \bar{x})^2}{(n-1)\sum_{r=1}^{n} w_r}}$$

Full method

In fact, the sample in Great Britain is a multi-stage, stratified, random sample described further in Appendix A. First a sample of areas, the Primary Sampling Units (PSUs), is drawn from an ordered list. Then within each PSU a random sample of households is drawn. In Northern Ireland, however, the sample is drawn in a single stage and there is no clustering. The results are also weighted for non-response and calibrated to match the population separately by sex, by 5-year age ranges and by region, as described in Appendix F.

The method for calculating complex standard errors for the weighted estimates used on this survey is quite complex. First, we apply methods that take account of the clustering, stratification and differential sampling (and initial nonresponse weights) used in the design. Then we modify these to allow for the calibration weighting used on the survey. The exact formulae also depend on whether we are estimating standard errors for an estimated total or a mean or proportion. Here we outline the method for a total.

Consecutive PSUs in the ordered list are first grouped up into pairs, or triples at the end of a regional stratum. The standard error of a weighted total is estimated by:

$$sedes = \sqrt{\sum_h \frac{k_h}{k_h - 1} \sum_i (x_{hi} - \bar{x}_h)^2}$$

where the h denotes the stratum (PSU pairs or triples), k_h is the number of PSUs in the stratum h (either 2 or 3), the x_{hi} is the weighted total in PSU i and the $\bar{x}_h$ is the mean of these totals in stratum h. Further details of this method of estimating sampling errors are described in *A sampling Errors Manual* (B Butcher and D Elliot, ONS 1987).

The effect of the calibration weighting is calculated using a jackknife linearisation estimator. It uses the formula given above but with each household's expenditure, x_r ,replaced by a residual from a linear regression of expenditure on the number of people in each household in each of the region and age by sex categories used in the weighting. Details are available from the editor.

The formulae have been expressed in terms of expenditures on a particular item, but of course they can also be applied to expenditures on groups of items, commodity groups and incomes from particular sources.

Appendix D
Definitions

Major changes in definitions since 1991 are described in Appendix E. Changes made between 1980 and 1990 are summarised in Appendix E of Family Spending 1994-95. For earlier changes see Annex 5 of Family Expenditure Survey 1980.

Contents Page

Household 174
- Retired households 174
- Household reference person 174
- Members of household 175
- Household composition 175
- Adult 175
- Children 175
- Spenders 176

Economically active 176
Economically inactive 176
NS-SEC 177
Regions 177
Urban/rural areas 177
Expenditure 177
- Goods supplied from a household's own shop or farm 178
- Hire purchase, credit sales agreements and loans 178
- Club payments 178
- Credit card transactions 179
- Income tax 179
- Rented dwellings 179
- Rent-free dwellings 179
- Owner-occupied dwellings 180
- Second-hand goods and part-exchange transactions 180
- Business expenses 180

Income 180
- Wages and salaries of employees 181
- Income from self-employment 182
- Income from investment 182
- Social security benefits 182

Quantiles 182
Income headings 183
Region diagram 185

Household

A household comprises one person or a group of people who have the accommodation as their only or main residence and (for a group):

share the living accommodation, that is a living room or sitting room
or share meals together or have common housekeeping

Resident domestic servants are included. The members of a household are not necessarily related by blood or marriage. As the survey covers only private households, people living in hostels, hotels, boarding houses or institutions are excluded. Households are not excluded if some or all members are not British subjects, but no attempt is made to obtain information from households containing members of the diplomatic service of another country or members of the United States armed forces. Nor are attempts made to obtain information from Roman Catholic priests living in accommodation provided by the parish church.

Retired households

Retired households are those where the household reference person is retired. The household reference person is defined as retired if 65 years of age or more and male or 60 years of age or more and female, and economically inactive. Hence if, for example, a male household reference person is over 65 years of age, but working part-time or waiting to take up a part-time job, this household would not be classified as a retired household. For analysis purposes two categories are used in this report:

a. "A retired household mainly dependent upon state pensions" is one in which at least three quarters of the total income of the household is derived from national insurance retirement and similar pensions, including housing and other benefits paid in supplement to or instead of such pensions. The term "national insurance retirement and similar pensions" includes national insurance disablement and war disability pensions, and income support in conjunction with these disability payments.

b. "Other retired households" are retired households which do not fulfil the income conditions of "retired household mainly dependent upon state pensions" because more than a quarter of the household's income derives from occupational retirement pensions and/or income from investments, annuities etc.

Household reference person (HRP)

From 2001-02, the concept of household reference person (HRP) was adopted on all government-sponsored surveys, in place of head of household. The household reference person is the householder, i.e. the person who:

a. owns the household accommodation, or
b. is legally responsible for the rent of the accommodation, or
c. has the household accommodation as an emolument or perquisite, or
d. has the household accommodation by virtue of some relationship to the owner who is not a member of the household.

If there are joint householders the household reference person will be the one with the higher income. If the income is the same, then the eldest householder is taken.

A key difference between household reference person and head of household is that the household reference person must always be a householder, whereas the head of household was always the husband, who might not even be a householder himself.

Members of household

In most cases the members of co-operating households are easily identified as the people who satisfy the conditions in the definition of a household, above, and are present during the record-keeping period. However difficulties of definition arise where people are temporarily away from the household or else spend their time between two residences. The following rules apply in deciding whether or not such persons are members of the household:

a. married persons living and working away from home for any period are included as members provided they consider the sampled address to be their main residence; in general, other people (e.g. relatives, friends, boarders) who are either temporarily absent or who spend their time between the sampled address and another address, are included as members if they consider the sampled address to be their main residence. However, there are exceptions which override the subjective main residence rule:

    i. Children under 16 away at school are included as members;

    ii. Older persons receiving education away from home, including children aged 16 and 17, are excluded unless they are at home for all or most of the record-keeping period.

    iii. Visitors staying temporarily with the household and others who have been in the household for only a short time are treated as members -provided they will be staying with the household for at least one month from the start of record-keeping.

Household composition

A consequence of these definitions is that household compositions quoted in this report include some households where certain members are temporarily absent. For example, "one adult and children" households will contain a few households where one parent is temporarily away from home.

Adult

In the report, persons who have reached the age of 18 or who are married are classed as adults.

Children

In the report, persons who are under 18 years of age and unmarried are classed as children.

However, in the definition of clothing, clothing for persons aged 16 years and over is classified as clothing for men and women; clothing for those aged five but under 16 as clothing for boys and girls; and clothing for those under five as babies clothing.

Spenders

Members of households who are aged 16 or more, excluding those who for special reasons are not capable of keeping diary record-books, are described as spenders.

Economically active

These are persons aged 16 or over who fall into the following categories:

a. *Employees at work* - those who at the time of interview were working full-time or part-time as employees or were away from work on holiday. Part-time work is defined as normally working 30 hours a week or less (excluding meal breaks) including regularly worked overtime.

b. *Employees temporarily away from work* - those who at the time of interview had a job but were absent because of illness or accident, temporary lay-off, strike etc.

c. *Government supported training schemes* - those participating in government programmes and schemes who in the course of their participation receive training, such as Employment Training, including those who are also employees in employment.

d. *Self-employed* - those who at the time of interview said they were self-employed.

e. *Unemployed* - those who at time of interview were out of employment, and have sought work within the last four weeks and were available to start work within two weeks, or were waiting to start a job already obtained.

f. *Unpaid family workers* - those working unpaid for their own or a relative's business. In this report, unpaid family workers are included under economically inactive in analyses by economic status (tables 3.1 and 9.1) because insufficient information is available to assign them to an economic status group.

Economically inactive

a. *Retired* - persons who have reached national insurance retirement age (60 and over for women, 65 and over for men) and are not economically active.

b. *Unoccupied* - persons under national insurance retirement age who are not working, nor actively seeking work. This category includes certain self-employed persons such as mail order agents and baby-sitters who are not classified as economically active.

In this report, unpaid family workers are classified as economically inactive in analyses by economic status, although they are economically active by definition. This is because insufficient information is available to assign them to an economic status group.

National Statistics Socio-economic classification (NS-SEC)

From 2001, the National Statistics Socio-economic classification (NS-SEC) was adopted for all official surveys, in place of Social Class based on Occupation and Socio-economic group. NS-SEC is itself based on the Standard Occupational Classification 2000 (SOC2000) and details of employment status. Although NS-SEC is an occupationally based classification, there are procedures for classifying those not in work. The main categories used for analysis in Family Spending are:

1 Higher managerial and professional occupations, sub-divided into:
  1.1 Large employers and higher managerial occupations
  1.2 Higher professional occupations
2 Lower managerial and professional occupations
3 Intermediate occupations
4 Small employers and own account workers
5 Lower supervisory and technical occupations
6 Semi-routine occupations
7 Routine occupations
8 Never worked and long-term unemployed

The long-term unemployed, which fall into a separate category, are defined as those unemployed and seeking work for 12 months or more. Members of the armed forces, who were assigned to a separate category in Social Class, are included within the NS-SEC classification. Residual groups that remain unclassified include students and those with inadequately described occupations. For the purposes of Family Spending, retired individuals are not assigned an NS-SEC category.

Regions

These are the Government Office Regions as formed in 1994. See Appendix D for more details.

Urban and rural areas

This classification is based on the population of the continuous built-up areas, irrespective of administrative boundaries derived by the Department for Transport, Local Government and the Regions (DTLR) based on the 1991 Census. Note that the metropolitan built-up areas are not the same as the metropolitan administrative districts. They exclude any rural areas within the metropolitan districts and include any built up areas adjoining them.

Expenditure

Any definition of expenditure is to some extent arbitrary, and the inclusion of certain types of payment is a matter of convenience or convention depending on the purpose for which the information is to be used. In the tables in this report, total expenditure represents current expenditure on goods and services. Total expenditure, defined in this way, excludes those recorded payments which are really savings or investments (e.g. purchases of national savings certificates, life assurance premiums, contributions to pension funds). Similarly, income tax payments, national insurance contributions, mortgage capital repayments and other payments for major additions to dwellings are excluded. Expenditure data are collected in the diary record-book and in the household schedule. Informants are asked to record in the diary any payments made during the 14 days of record-keeping, whether or not the goods or services paid for have been received. Certain types of expenditure which are usually regular though infrequent, such as insurance, licences and season tickets, and the periods to which they relate, are recorded in the household schedule as well as regular payments such as utility bills.

The cash purchase of motor vehicles is also entered in the household schedule. In addition, expenditure on some items purchased infrequently (thereby being subject to high sampling errors) has been recorded in the household schedule using a retrospective recall period of either three or 12 months. These items include carpets, furniture, holidays and some housing costs. In order to avoid duplication, all payments shown in the diary record-book which relate to items listed in the household or income schedules are omitted in the analysis of the data irrespective of whether there is a corresponding entry on the latter schedules. Amounts paid in respect of periods longer than a week are converted to weekly values.

Expenditure tables in this report show the 12 main commodity groups of spending and these are broken down into items which are numbered hierarchically (see Appendix E which details a major change to the coding frame used from 2001-02). Table 7.1 shows a further breakdown in the items themselves into components which can be separately identified. The items are numbered as in the main expenditure tables and against each item or component are shown the average weekly household expenditure and percentage standard error.

Qualifications which apply to this concept of expenditure are described in the following paragraphs:

a. *Goods supplied from a household's own shop or farm*
Spenders are asked to record and give the value of goods obtained from their own shop or farm, even if the goods are withdrawn from stock for personal use without payment. The value is included as expenditure.

b. *Hire purchase and credit sales agreements, and transactions financed by loans repaid by instalments*
Expenditure on transactions under hire purchase or credit sales agreements, or financed by loans repaid by instalments, consists of all instalments which are still being paid at the date of interview, together with down payments on commodities acquired within the preceding three months. These two components (divided by the periods covered) provide the weekly averages which are included in the expenditure on the separate items given in the tables in this report.

c. *Club payments and budget account payments, instalments through mail order firms and similar forms of credit transaction*
When goods are purchased by forms of credit other than hire purchase and credit sales agreement, the expenditure on them may be estimated either from the amount of the instalment which is paid or from the value of the goods which are acquired. Since the particular commodities to which the instalment relates may not be known, details of goods ordered through clubs, etc. during the month prior to the date of interview are recorded in the household schedule. The weekly equivalent of the value of the goods is included in the

expenditure on the separate items given in the tables in this report. This procedure has the advantage of enabling club transactions to be related to specific articles. Although payments into clubs, etc. are shown in the diary record-book, these entries are excluded from expenditure estimates.

d. *Credit card transactions*

From 1988 purchases made by credit card or charge card have been recorded in the survey on an *acquisition* basis rather than the formerly used payment basis. Thus, if a spender acquired an item (by use of credit/charge card) during the two week survey period, the value of the item would be included as part of expenditure in that period whether or not any payment was made in this period to the credit card account. Payments made to the card account are ignored. However any payment of credit/charge card *interest* is included in expenditure if made in the two week period.

e. *Income Tax*

Amounts of income tax deducted under the PAYE scheme or paid directly by those who are employers or self-employed are recorded (together with information about tax refunds). For employers and the self-employed the amounts comprise the actual payments made in the previous twelve months and may not correspond to the tax due on the income arising in that period, e.g. if no tax has been paid but is due or if tax payments cover more than one financial year. However, the amounts of tax deducted at source from some of the items which appear in the Income Schedule are not directly available. Estimates of the tax paid on bank and building society interest and amounts deducted from dividends on stocks and shares are therefore made by applying the appropriate rates of tax. In the case of income tax paid at source on pensions and annuities, similar adjustments are made. These estimates mainly affect the relatively few households with high incomes from interest and dividends, and households including someone receiving a pension from previous employment.

f. *Rented dwellings*

Housing expenditure is taken as the sum of expenditure on rent, rates, council tax, water rates etc. For local authority tenants the expenditure is gross rent less any rebate (including rebate received in the form of housing benefit), and for other tenants gross rent less any rent allowance received under statutory schemes including the Housing Benefit Scheme. Rebate on Council Tax or rates (Northern Ireland) is deducted from expenditure on Council Tax or rates. Receipts from sub-letting part of the dwelling are not deducted from housing costs but appear (net of the expenses of the sub-letting) as investment income: see page 149. Average payments by households renting accommodation for repairs, maintenance and decorations are shown separately in the estimates of expenditure by such households in table 4.10 which gives housing expenditure by tenure type. Accommodation rented from a housing association is shown separately.

g. *Rent-free dwellings*

Rent-free dwellings are those owned by someone outside the household and where either no rent is charged or the rent is paid by someone outside the household. Households whose rent is paid directly to the landlord by the DSS do not live rent-free. Payments for Council Tax, water rates etc., are regarded as the cost of housing. Rebate on rates(Northern Ireland)/ Council Tax/water rates(Scotland) (including rebate received in the form of housing benefit), is deducted from expenditure on rates/Council Tax/water rates. Receipts from sub-letting part of the dwelling are not deducted from housing costs but appear (net of the expenses of the sub-letting) as investment income.

h. *Owner-occupied dwellings*
In the EFS payments for water rates, ground rent, fuel, maintenance and repair of the dwelling, and other miscellaneous services related to the dwelling etc., are regarded as the cost of housing. Receipts from letting part of the dwelling are not deducted from housing costs but appear (net of the expenses of the letting) as investment income. Mortgage capital repayments and amounts paid for the outright purchase of the dwelling or for major structural alterations are not included as housing expenditure, but are entered under "Other items recorded", as are Council Tax, rates (Northern Ireland), and mortgage interest payments. Structural insurance is included in Miscellaneous goods and services.

i. *Second-hand goods and part-exchange transactions*
The survey expenditure data are based on information about actual payments and therefore include payments for second-hand goods and part-exchange transactions. New payments only are included for part-exchange transactions, i.e. the costs of the goods obtained less the amounts allowed for the goods which are traded in. Receipts for goods sold or traded in are not included in income.

j. *Business expenses*
The survey covers only private households and is concerned with payments made by members of households as private individuals. Spenders are asked to state whether expenditure which has been recorded on the schedules includes amounts which will be refunded as expenses from a business or organisation or which will be entered as business expenses for income tax purposes, e.g. rent, telephone charges, travelling expenses, meals out. Any such amounts are deducted from the recorded expenditure.

Income

The standard concept of income in the survey is, as far as possible, that of gross weekly cash income current at the time of interview, i.e. before the deduction of income tax actually paid, national insurance contributions and other deductions at source. However, for a few tables a concept of disposable income is used, defined as gross weekly cash income less the statutory deductions and payments of income tax (taking refunds into account) and national insurance contributions. Some other analyses of EFS data use "equivalisation" of incomes - i.e. adjustment of household income to allow for the different size and composition of each household. Equivalisation is not used in this volume. The cash levels of certain items of income (and expenditure) recorded in the survey by households receiving supplementary benefit were affected by the Housing Benefit Scheme introduced in stages from November 1982. From 1984 housing expenditure is given on a strictly net basis and all rent/council tax rebates and allowances and housing benefit are excluded from gross income.

Although information about most types of income is obtained on a current basis, some data, principally income from investment and from self-employment, are estimated over a twelve-month period.

The following are excluded from the assessment of income:

a. money received by one member of the household from another (e.g. housekeeping money, dress allowance, children's pocket money) other than wages paid to resident domestic servants;

b. withdrawals of savings, receipts from maturing insurance policies, proceeds from sale of financial and other assets (e.g. houses, cars, furniture, etc.), winnings from betting, lump-sum gratuities and windfalls such as legacies;

c. the value of educational grants and scholarships not paid in cash;

d. the value of income in kind, including the value of goods received free and the abatement in cost of goods received at reduced prices, and of bills paid by someone who is not a member of the household;

e. loans and money received in repayment of loans.

Details are obtained of the income of each member of the household. The income of the household is taken to be the sum of the incomes of all its members. The information does not relate to a common or a fixed time period. Items recorded for periods greater than a week are converted to a weekly value.

Particular points relating to some components of income are as follows:

a. *Wages and salaries of employees*
The normal gross wages or salaries of employees are taken to be their earnings. These are calculated by adding to the normal "take home" pay amounts deducted at source, such as income tax payments, national insurance contributions and other deductions, e.g. payments into firm social clubs, superannuation schemes, works transport, benevolent funds etc. Employees are asked to give the earnings actually received including bonuses and commission the last time payment was made and, if different, the amount usually received. It is the amount usually received which is regarded as the normal take-home pay. Additions are made so as to include in normal earnings the value of occasional payments, such as bonuses or commissions received quarterly or annually. One of the principal objects in obtaining data on income is to enable expenditure to be classified in ranges of normal income. Average household expenditure is likely to be based on the long-term expectations of the various members of the household as to their incomes rather than be altered by short-term changes affecting individuals. Hence if employees have been away from work without pay for 13 weeks or less they are regarded as continuing to receive their normal earnings instead of social security benefits, such as unemployment or sickness benefit, that they may be receiving. Otherwise, normal earnings are disregarded and current short-term social security benefits taken instead. Wages and salaries include any earnings from subsidiary employment as an employee and the earnings of HM Forces.

b. *Income from self-employment*

Income from self-employment covers any personal income from employment other than as an employee; for example, as a sole trader, professional or other person working on his own account or in partnership, including subsidiary work on his own account by a person whose main job is as an employee. It is measured from estimates of income or trading profits, after deduction of business expenses but before deduction of tax, over the most recent twelve-month period for which figures can be given. Should either a loss have been made or no profit, income would be taken as the amounts drawn from the business for own use or as any other income received from the job or business. Persons working as mail order agents or baby-sitters, with no other employment, have been classified as unoccupied rather than as self-employed, and the earnings involved have been classified as earnings from "other sources" rather than self-employment income.

c. *Income from investment*

Income from investments or from property, other than that in which the household is residing, is the amount received during the twelve months immediately prior to the date of the initial interview. It includes receipts from sub-letting part of the dwelling (net of the expenses of the sub-letting). If income tax has been deducted at source the gross amount is estimated by applying a conversion factor during processing.

d. *Social security benefits*

Income from social security benefits does not include the short-term payments such as unemployment or sickness benefit received by an employee who has been away from work for 13 weeks or less, and who is therefore regarded as continuing to receive his normal earnings as described on page 151.

Quantiles

The quantiles of a distribution, e.g. of household expenditure or income, divide it into a number of equal parts; each of which contains the same number of households.

For example, the median of a distribution divides it into two equal parts, so that half the households in a distribution of household income will have income more than the median, and the other half will have income less than the median. Similarly, quartiles, quintiles and deciles divide the distribution into four, five and ten equal parts respectively.

Most of the analysis in Family Spending is done in terms of quintile groups and decile groups.

In the calculation of quantiles for this report, zero values are counted as part of the distribution.

Income headings

Headings used for identifying 2002-03 income information

	Source of income	
References in tables	Components separately identified	Explanatory notes
a. Wages and salaries	Normal "take-home" pay from main employment "Take-home" pay from subsidiary employment Employees' income tax deduction Employees' National Insurance contribution Superannuation contributions deducted from pay Other deductions	(i) In the calculation of house hold income in this report, where an employee has been away from work without pay for 13 weeks or less his normal wage or salary has been used in estimating his total income instead of social security benefits, such as unemployment or sickness benefits that he may have received. Otherwise such benefits are used in estimating total income (see notes at reference e) (ii) Normal income from wages and salaries is estimated by adding to the normal "take-home" pay deductions made at source last time paid, together with the weekly value of occasional additions to wages and salaries (see page 148). (iii) The components of wages and salaries for which figures are separately available amount in total to the normal earnings of employees, regardless of the operation of the 13 week rule in note (i) above. Thus the sum of the components listed here does not in general equal the wages and salaries figure in tables of this report.
b. Self-employment	Income from business or profession, including subsidiary self-employment expenses but before deduction of tax	The earnings or profits of a trade or profession, after deduction of business
c. Investments	Interest on building society shares and deposits Interest on bank deposits and savings accounts including National Savings Bank Interest on ISAs Interest on TESSAs Interest on Gilt-edged stock and War Loans Interest and dividends from stocks, shares, bonds, trusts, PEPs, debentures and other securities Rent or income from property, after deducting expenses but inclusive of income tax (including receipts from letting or sub-letting part of own residence, net of the expenses of the letting or sub-letting). Other unearned Income	

d. Annuities and pensions,other than social security	Annuities and income from trust or covenant Pensions from previous employers Personal pensions	
e. Social security benefits	Child benefit Guardian's allowance Invalid care allowance Retirement pension (National Insurance) or old person's pension Widow's pension or widowed mother's allowance (NI) War disablement pension or war widow's pension Severe disablement allowance Disabled person's tax credit Care component of disability living allowance Mobility component of disability living allowance Attendance allowance Job seekers allowance, contributions based Job seekers allowance, income based Income support Working families tax credit Incapacity benefit Statutory sick pay (from employer) Industrial injury disablement benefit Maternity allowance Statutory maternity pay Any other benefit including lump sums and grants Social security benefits excluded from income calculation by 13 week rule	I. The calculation of household income in this report takes account of the 13 week rule described at reference a, note (i) ii. The components of social security benefits for which figures are separately available amount in total to the benefits received in the week before interview. That is to say, they include amounts that are discounted from the total by the operation of the 13 week rule in note i. Thus the sum of the components listed here differs from the total of social security benefits used in the income tables of this report. iii Housing Benefit is treated as a reduction in housing costs and not as income
f. Other sources	Married person's allowance from husband/wife temporarily away from home Alimony or separation allowances; allowances for foster children, allowances from members of the Armed Forces or Merchant Navy, or any other money from friends or relatives, other than husband outside the household Benefits from trade unions, friendly societies etc., other than pensions Value of meal vouchers Earnings from intermittent or casual work over twelve months, not included in a or b above Student loans and money scholarships received by persons aged 16 and over and aged under 16. Other income of children under 16	e.g. from spare-time jobs or income from trusts or investments

STANDARD STATISTICAL REGION	COUNTY*	GOVERNMENT OFFICE REGION
NORTH	Cleveland*	NORTH EAST
NORTH	Durham	NORTH EAST
NORTH	Northumberland	NORTH EAST
NORTH	Tyne and Wear	NORTH EAST
NORTH	Cumbria	NORTH WEST
NORTH WEST	Cheshire	NORTH WEST
NORTH WEST	Greater Manchester	NORTH WEST
NORTH WEST	Lancashire	NORTH WEST
NORTH WEST	Merseyside	NORTH WEST
YORKSHIRE AND HUMBERSIDE	Humberside*	YORKSHIRE AND THE HUMBER
YORKSHIRE AND HUMBERSIDE	North Yorkshire*	YORKSHIRE AND THE HUMBER
YORKSHIRE AND HUMBERSIDE	South Yorkshire	YORKSHIRE AND THE HUMBER
YORKSHIRE AND HUMBERSIDE	West Yorkshire	YORKSHIRE AND THE HUMBER
EAST MIDLANDS	Derbyshire	EAST MIDLANDS
EAST MIDLANDS	Leicestershire	EAST MIDLANDS
EAST MIDLANDS	Lincolnshire	EAST MIDLANDS
EAST MIDLANDS	Northamptonshire	EAST MIDLANDS
EAST MIDLANDS	Nottinghamshire	EAST MIDLANDS
WEST MIDLANDS	Hereford and Worcester	WEST MIDLANDS
WEST MIDLANDS	Shropshire	WEST MIDLANDS
WEST MIDLANDS	Staffordshire	WEST MIDLANDS
WEST MIDLANDS	Warwickshire	WEST MIDLANDS
WEST MIDLANDS	West Midlands	WEST MIDLANDS
EAST ANGLIA	Cambridgeshire	EAST OF ENGLAND
EAST ANGLIA	Norfolk	EAST OF ENGLAND
EAST ANGLIA	Suffolk	EAST OF ENGLAND
SOUTH EAST	Bedfordshire	EAST OF ENGLAND
SOUTH EAST	Essex	EAST OF ENGLAND
SOUTH EAST	Hertfordshire	EAST OF ENGLAND
SOUTH EAST	Greater London	LONDON
SOUTH EAST	Berkshire	SOUTH EAST
SOUTH EAST	Buckinghamshire	SOUTH EAST
SOUTH EAST	East Sussex	SOUTH EAST
SOUTH EAST	Hampshire	SOUTH EAST
SOUTH EAST	Isle of Wight	SOUTH EAST
SOUTH EAST	Kent	SOUTH EAST
SOUTH EAST	Oxfordshire	SOUTH EAST
SOUTH EAST	Surrey	SOUTH EAST
SOUTH EAST	West Sussex	SOUTH EAST
SOUTH WEST	Avon*	SOUTH WEST
SOUTH WEST	Cornwall	SOUTH WEST
SOUTH WEST	Devon	SOUTH WEST
SOUTH WEST	Dorset	SOUTH WEST
SOUTH WEST	Gloucestershire	SOUTH WEST
SOUTH WEST	Somerset	SOUTH WEST
SOUTH WEST	Wiltshire	SOUTH WEST

* Counties prior to local government reorganisation

Appendix E
Changes in definitions, 1991 to 2002-03

1991
No significant changes.

1992
Housing – Imputed rent for owner occupiers and households in rent-free accommodation was discontinued. For owner occupiers this had been the rent they would have had pay themselves to live in the property they own, and for households in rent-free accommodation it was the rent they would normally have had to pay. Up to 1990 these amounts were counted both as income and as a housing cost. Mortgage interest payments were counted as a housing cost for the first time in 1991.

1993
Council Tax - Council Tax was introduced to replace the Community Charge in Great Britain from April 1993.

1994-95
New expenditure items - The definition of expenditure was extended to include two items previously shown under "other payments recorded". These were:

gambling payments;
mortgage protection premiums.

Expenditure classifications - A new classification system for expenditures was introduced in April 1994. The system is hierarchical and allows more detail to be preserved than the previous system. New categories of expenditure were introduced and are shown in detail in table 7.1. The 14 main groups of expenditure were retained, but there were some changes in the content of these groups.

Gambling Payments - data on gambling expenditure and winnings are collected in the expenditure diary. Previously these were excluded from the definition of household expenditure used in the FES. The data are shown as memoranda items under the heading "Other payments recorded" on both gross and net bases. The net basis corresponds approximately to the treatment of gambling in the National Accounts. The introduction of the National Lottery stimulated a reconsideration of this treatment. From April 1994, (gross) gambling payments have been included as expenditure in "Leisure Services". Gambling winnings continued to be noted as a memorandum item under "Other items recorded". They are treated as windfall income. They do not form a part of normal household income, nor are they subtracted from gross gambling payments. This treatment is in line with the PRODCOM classification of the Statistical Office of the European Communities (SOEC) for expenditure in household budget surveys.

1995-96
Geographical overage - The FES geographical coverage was extended to mainland Scotland north of the Caledonian Canal.

Under 16s diaries - Two week expenditure diaries for 7-15 year olds were introduced following three feasibility pilot studies which found that children of that age group were able to cope with the task of keeping a two week expenditure record. Children are asked to record everything they buy with their own money but to exclude items bought with other people's money. Purchases are coded according to the same coding categories as adult diaries except for meals and snacks away from home which are coded as school meals, hot meals and snacks, and cold meals and snacks. Children who keep a diary are given a £5 incentive payment. A refusal to keep an under 16's diary does not invalidate the household from inclusion in the survey.

Pocket money given to children is still recorded separately in adult diaries; and money paid by adults for school meals and school travel is recorded in the Household Questionnaire. Double counting is eliminated at the processing stage.

Tables in Family Spending reports did not include the information from the children's diaries until the 1998-99 report. Appendix F in the 1998-99 and 1999-2000 reports show what difference the inclusion made.

1996-97

Self-employment - The way in which information about income from self-employment is collected was substantially revised in 1996-97 following various tests and pilot studies. The quality of such data was increased but this may have lead to a discontinuity. Full details are shown the Income Questionnaire, available from the address in the introduction.

Cable/satellite television - Information on cable and satellite subscriptions is now collected from the household questionnaire rather than from the diary, leading to more respondents reporting this expenditure.

Mobile phones - Expenditure on mobile phones was previously collected through the diary. From 1996/97 this has been included in the questionnaire.

Job seekers allowance (JSA) - Introduced in October 1996 as a replacement for Unemployment Benefit and any Income Support associated with the payment of Unemployment Benefit. Receipt of JSA is collected with NI Unemployment Benefit and with Income Support. In both cases the number of weeks a respondent has been in receipt of these benefits is taken as the number of weeks receiving JSA in the last 12 months and before that period the number of weeks receiving Unemployment Benefit/Income Support.

Retrospective recall - The period over which information is requested has been extended from 3 to 12 months for vehicle purchase and sale. Information on the purchase of car and motorcycle spare parts is no longer collected by retrospective recall. Instead expenditure on these items is collected through the diary.

State benefits - The lists of benefits specifically asked about was reviewed in 1996/97. See the Income Questionnaire for more information.

Sample stratifiers - New stratifiers were introduced in 1996/97 based on standard regions, socio-economic group and car ownership.

Government Office Regions - Regional analyses are now presented using the Government Office Regions (GORs) formed in 1994. Previously all regional analyses used Standard Statistical Regions (SSRs). For more information see Appendix F.

1997-98

Bank/Building society service charges - Collection of information on service charges levied by banks has been extended to include building societies.

Payments from unemployment/redundancy insurances - Information is now collected on payments received from private unemployment and redundancy insurance policies. This information is then incorporated into the calculation of income from other sources.

Retired households - The definition of retired households has been amended to exclude households where the head of the household is economically active.

Rent-free tenure - The definition of rent-free tenure has been amended to include those households for which someone outside the household, except an employer or an organisation, is paying a rent or mortgage on behalf of the household.

National Lottery - From February 1997, expenditure on National lottery tickets was collected as three separate items: tickets for the Wednesday draw only, tickets for the Saturday draw only and tickets for both draws.

1998-99

Children's income – Three new expenditure codes were introduced: pocket money to children; money given to children for specific purposes and cash gifts to children. These replaced a single code covering all three categories.

Main job and last paid job – Harmonised questions were adopted.

1999-2000

Disabled Persons Tax Credit replaced Disability Working Allowance and *Working Families Tax Credit* replaced Family Credit from October 1999.

2000-01

Household definition – the definition was changed to the harmonised definition which has been in use in the Census and nearly all other government household surveys since 1981. The effect is to group together into a single household some people who would have been allocated to separate households on the previous definition. The effect is fairly small but not negligible.

Up to 1999-2000 the FES definition was based on the pre-1981 Census definition and required members to share eating and budgeting arrangements as well as shared living accommodation. The definition of a household was:

One person or a group of people who have the accommodation as their only or main residence and (for a group)

share the living accommodation, that is a living or sitting room

and share meals together (or have common housekeeping).

The harmonised definition is less restrictive:

One person or a group of people who have the accommodation as their only or main residence and (for a group)

share the living accommodation, that is a living or sitting room

or share meals together or have common housekeeping.

The effect of the change is probably to increase average household size by 0.6 per cent.

Question reductions - A thorough review of the questionnaire showed that a number of questions were no longer needed by government users. These were cut from the 2000-01 survey to reduce the burden on respondents. The reduction was fairly small but it did make the interview flow better. All the questions needed for a complete record of expenditure and income were retained.

Redesigned diary - The diary was redesigned to be easier for respondent to keep and to look cleaner. The main change of substance was to delete the column for recording whether each item was purchased by credit, charge or shop card.

Ending of MIRAS - Tax relief on interest on loans for house purchase was abolished from April 2000. Questions related to MIRAS were therefore dropped. They included some that were needed to estimate the amount if the respondent did not know it. A number were retained for other purposes, however, such as the amount of the loan still outstanding which is still asked for households paying a reduced rate of interest because one of them works for the lender.

2001-02

Household reference person – this replaced the previous concept of head of household. The household reference person is the householder, i.e. the person who:

a. owns the household accommodation, or
b. is legally responsible for the rent of the accommodation, or
c. has the household accommodation as an emolument or perquisite, or
d. has the household accommodation by virtue of some relationship to the owner who is not a member of the household.

If there are joint householders the household reference person is the one with the higher income. If the income is the same, then the eldest householder is taken.

A key difference between household reference person and head of household is that the household reference person must always be a householder, whereas the head of household was always the husband, who might not even be a householder himself.

National Statistics Socio-economic classification (NS-SEC) **–** the National Statistics Socio-economic classification (NS-SEC) was adopted for all official surveys, in place of Social Class based on Occupation and Socio-economic group. NS-SEC is itself based on the Standard Occupational Classification 2000 (SOC2000) and details of employment status.

The long-term unemployed, which fall into a separate category, are defined as those unemployed and seeking work for 12 months or more. Members of the armed forces, who were assigned to a separate category in Social Class, are included within the NS-SEC classification. Residual groups that remain unclassified include students and those with inadequately described occupations.

COICOP – From 2001-02, the **C**lassification **Of** **I**ndividual **CO**nsumption by **P**urpose (COICOP/HBS, referred to as COICOP in this volume) was introduced as a new coding frame for expenditure items. COICOP has been adapted to the needs of Household Budget Surveys (HBS) across the EU and, as a consequence, is compatible with similar classifications used in national accounts and consumer price indices. This allows the production of indicators which are comparable Europe-wide, such as the Harmonised Indices of Consumer Prices (computed for all goods as well as sub-categories such as food and transport). The main categorisation of spending used in this volume (namely twelve categories relating to food and non-alcoholic beverages; alcoholic beverages, tobacco and narcotics; clothing and footwear; housing, fuel and power; fhousehold goods and services; health, transport; communication; recreation and culture; education; restaurants and hotels; and miscellaneous goods and services) is only comparable between the two frames at a broad level. Table 6.1 in this volume has been produced by mapping COICOP to the FES 14 main categories. However the two frames are not comparable for any smaller categories, leading to a break in trends between 2000-01 and 2001-02 for any level of detail below the main 12-fold categorisation. A complete listing of COICOP and COICOP plus (an extra level of detail added by individual countries for their own needs) is available on request from the address in the introduction.

Proxy interviews – While questions about general household affairs are put to all household members or to a main household informant, questions about work and income are put to the individual members of the household. Where a member of the household is not present during the household interview, another member of the household (e.g. spouse) may be able to provide information about the absent person. The individual's interview is then identified as a *proxy* interview. From 2001/02, the EFS began accepting responses that contained a proxy interview.

Short income – From 2001-02, the EFS accepted responses from households that answered the short income section. This was designed for respondents who were reluctant to provide more detailed income information.

2002-03

Main shopper – At the launch of the EFS in April 2001, the respondent responsible for buying the household's main shopping was identified as the Main Diary Keeper. From 2002-03, this term has been replaced by the "Main Shopper".

The importance of the Main Shopper is to ensure that we have obtained information on the bulk of the shopping in the household. Without this person's co-operation we have insufficient information to use the other diaries kept by members of the household in a meaningful way. The main shopper must therefore complete a diary for the interview to qualify as a full or partial interview. Without their participation, the outcome will be a refusal no matter who else is willing to complete a diary.

Appendix F
Differential grossing

Since 1998-99 results have been based on data that have been grossed differentially (re-weighted) to reduce the effect of non-response bias. This appendix shows the effect on the 2002-03 results published in this report. The population weights are based on the latest population estimates from the 2001 census.

The grossing method used for the 2002-03 data is the same in principle as in previous years. As well as providing users with estimates of total spending by a single, agreed procedure, the grossing also re-weights the data. It is known from comparisons with the census (see the Appendix A section on reliability) that response rates are higher in some groups than others, leading to sampled households not being fully representative of the population as a whole. The aim of re-weighting is to compensate for this non-response bias by giving higher weights to households in the groups that are under-represented. An example of such an under-represented group is households with three or more adults and no children.

Method used to produce the weights

The weights are produced in two stages, the first of which uses results from the census-linked study of survey non-respondents (*Weighting the FES to compensate for non-response, Part 1: An investigation into census-based weighting schemes*, Foster 1994). A statistical analysis[1] was used to identify ten groups with very different response rates. A weight was then assigned to each of those groups, based on the inverse of the response rate for the group. A group with a low response rate is therefore given a high initial weight. ONS is currently analysing data from the 2001 Census; this will result in an update to the non-response weights applied to the EFS data.

The second stage adjusts the weights so that there is an exact match with population estimates, for males and females in different age groups and separately for regions. An important feature of the EFS grossing is that this is done by adjusting the factors for whole households, not by adjusting the factors for individuals. The population figures being matched exclude people who are not covered by the EFS, that is those in bed-and-breakfast accommodation, hostels, residential care homes and other institutions. A so called calibration method [2] is used in this stage to produce the weights.

The grossing procedure is carried out separately for each quarter of the survey. The main reason is that sample sizes vary from quarter to quarter more than in the past. This is the result of re-issuing addresses where there had been a non-contact or a refusal to a new interviewer after an interval of a few months, so that there are more interviews in the later quarters of the year than in the first quarter. Spending patterns are seasonal and quarterly grossing counteracts any bias from the uneven spread of interviews through the year. Quarterly grossing results in small sample numbers in some of the age group/sex categories previously used in the grossing and they have been widened slightly to avoid this.

1 Chi-squared Auomatic Interaction Detector

2 Implemented by the CALMAR software package

The overall effect of differential grossing

Table F1 shows the effect of differential grossing (weighting) on the 2002-03 EFS data.

Weighting increased the estimate of total average expenditure by £7.20 a week, that is by 2.1 per cent. It had the largest impact on average weekly expenditure on housing, fuel and power, increasing the estimate of expenditure by 4.7 per cent; on restaurants and hotels, increasing the expenditure estimate by 4.3 per cent; and transport costs by 3.7 per cent. It reduced the estimate of spending on health by 2.0 per cent. Weighting also increased the estimates of average income, by £8.70 a week (2.0 per cent) for disposable household income and by £12.40 a week (2.3 per cent) for gross household income, which is the income used in most tables in the report.

Table F1
The effect of weighting on expenditure

Commodity or service	Average weekly household expenditure Unweighted	Weighted as published	Absolute difference	*Percentage difference*
All expenditure groups	**341.10**	**348.30**	**7.20**	*2.1*
Food and non-alcoholic drinks	42.90	42.70	-0.10	*-0.3*
Alcoholic drink and tobacco	11.40	11.40	0.00	*-0.1*
Clothing and footwear	22.50	22.30	-0.20	*-1.1*
Housing, fuel and power	35.20	36.90	1.70	*4.7*
Household goods and services	30.10	30.20	0.10	*0.5*
Health	4.90	4.80	-0.10	*-2.0*
Transport	57.10	59.20	2.10	*3.7*
Communication	10.30	10.60	0.30	*2.8*
Recreation and culture	54.70	56.40	1.70	*3.0*
Education	5.10	5.20	0.10	*2.6*
Restaurants and hotels	34.00	35.40	1.50	*4.3*
Miscellaneous	32.90	33.10	0.20	*0.7*
Weekly household income:				
Disposable	444.70	453.40	8.70	*2.0*
Gross	539.90	552.30	12.40	*2.3*

Re-weighting also has an effect on the variance of estimates. In an analysis on the 1999-2000 data weighting increased variance slightly for some items and reduced for others. Overall the effect was to reduce variance slightly.

Further information

Further information is available on the method used to produce the weights from the address given in the introduction.

Appendix G

Index to tables in reports on the Family Expenditure Survey 1994-95 to 2000-01 and the Expenditure and Food Survey 2001-02 to 2002-03

		Table numbers in reports for							
2002-03 tables		2001-02[1]	2000-01	1999-2000	1998-99	1997-98	1996-97	1995-96	1994-95
1	**Expenditure by income**								
1.1	main items by gross income decile	1.1	1.1	1.1	1.1	1.1	1.1	1.1	1.1
1.2	percentage on main items by gross income decile	1.2	1.2	1.2	1.2	1.2	1.2	1.2	1.2
1.3	detailed expenditure by gross income decile	1.3	1.3	1.3	1.3	1.3	1.3	1.3	1.3
..	(housing expenditure in each tenure group)	-	-	-	-	-	1.4	1.4	1.4
1.4	main items by disposable income decile	1.4	1.4	1.4	-	-	-	-	-
1.5	percentage on main items by disposable income decile	1.5	1.5	1.5	-	-	-	-	-
2	**Expenditure by age and income**								
2.1	main items for all age groups of household reference person	2.1	2.9	-	-	-	-	-	-
..	main items for all age groups	-	2.1	2.1	2.1	2.1	2.1	2.1	2.1
2.2	main items as a percentage for all age groups	2.2	2.2	2.2	2.2	2.2	2.2	2.2	2.2
2.3	detailed expenditure for all age groups	2.3	2.3	2.3	2.3	2.3	2.3	2.3	2.3
2.4	aged under 30 by income	2.4	2.4	2.4	2.4	2.4	2.4	2.4	2.4
2.5	aged 30 and under 50 by income	2.5	2.5	2.5	2.5	2.5	2.5	2.5	2.5
2.6	aged 50 and under 65 by income	2.6	2.6	2.6	2.6	2.6	2.6	2.6	2.6
2.7	aged 65 and under 75 by income	2.7	2.7	2.7	2.7	2.7	2.7	2.7	2.7
2.8	aged 75 or over by income	2.8	2.8	2.8	2.8	2.8	2.8	2.8	2.8
3	**Expenditure by socio-economic characteristics**								
3.1	by economic activity status of household reference person	3.1	3.9	-	-	-	-	-	-
..	by economic activity status of head of household	-	3.1	3.1	3.1	3.1	3.1	3.1	3.1
..	by occupation	-	3.2	3.2	3.2	3.2	3.2	3.2	3.2
3.2	full-time employee by income	3.2	3.3	3.3	3.3	3.3	3.3	3.3	3.3
3.3	self-employed by income	3.3	3.4	3.4	3.4	3.4	3.4	3.4	3.4
..	by social class	-	3.5	3.5	3.5	3.5	3.5	3.5	3.5
3.4	by number of persons working	3.4	3.6	3.6	3.6	3.6	3.6	3.6	3.6
3.5	by age completed continuous full-time education	3.5	3.7	3.7	3.7	3.7	3.7	3.7	3.7
..	by occupation of household reference person	-	3.8	-	-	-	-	-	-
3.6	by socio-economic class of household reference person	3.6	-	-	-	-	-	-	-
4	**Expenditure by composition, income and tenure**								
4.1	expenditure by household composition	4.1	4.1	4.1	4.1	4.1	4.1	4.1	4.1
4.2	one adult retired households mainly dependent on state pensions	4.2	4.2	4.2	4.2	4.2	4.2	4.2	4.2
4.3	one adult retired households not mainly dependent on state pensions	4.3	4.3	4.3	4.3	4.3	4.3	4.3	4.3
4.4	one adult non-retired	4.4	4.4	4.4	4.4	4.4	4.4	4.4	4.4
4.5	one adult with children	4.5	4.5	4.5	4.5	4.5	4.5	4.5	4.5
4.6	two adults with children	4.6	4.6	4.6	4.6	4.6	4.6	4.6	4.6
4.7	one man one woman non-retired	4.7	4.7	4.7	4.7	4.7	4.7	4.7	4.7
4.8	one man one woman retired mainly dependent on state pensions	4.8	4.8	4.8	4.8	4.8	4.8	4.8	4.8

.. Tables do not appear in the 2002-03 report
1 Household reference person replaced head of household in 2001-02

	2002-03 tables	Table numbers in reports for							
		2001-02[1]	2000-01	1999-2000	1998-99	1997-98	1996-97	1995-96	1994-95
4.9	one man one woman retired not mainly dependent on state pensions	4.9	4.9	4.9	4.9	4.9	4.9	4.9	4.9
4.10	household expenditure by tenure	4.10	4.10	4.10	4.10	4.10	4.10	4.10	4.10
..	household expenditure by type of dwelling	-	-	-	4.11	4.11	4.11	4.11	4.11
5	**Expenditure by region[2,3]**								
5.1	main items of expenditure	5.1	5.1	5.1	5.1	5.1	5.1	5.1	5.1
5.2	main items as a percentage of expenditure	5.2	5.2	5.2	5.2	5.2	5.2	5.2	5.2
5.3	detailed expenditure	5.3	5.3	5.3	5.3	5.3	5.3	5.3	5.3
..	(housing expenditure in each tenure group)	-	-	-	-	-	5.4	5.4	5.4
..	expenditure by type of administrative area	-	5.4	5.4	5.4	5.4	5.5	5.5	5.5
5.4	expenditure by urban/rural areas (GB only)	5.4	5.5	-	-	-	-	-	-
6	**Trends in household expenditure**								
6.1	main items 1974 - 2000-01	6.1	6.1	6.1	-	-	-	-	-
6.2	as a percentage of total expenditure	6.2	6.2	6.2	6.1	6.1	6.1	6.1	6.1
..	by Region[3]	-	6.3	6.3	6.2	-	-	-	-
7	**Detailed expenditure and place of purchase**								
7.1	with full method standard errors	7.1	7.1	7.1	7.1	7.1	7.1	7.1	7.1
7.2	expenditure on alcoholic drink by type of premises	7.2	7.2	7.2	7.2	7.2	7.2	7.2	7.2
7.3	expenditure on food by place of purchase	7.3	7.3	7.3	7.3	7.3	7.3	7.3	7.3
..	expenditure on alcoholic drink by place of purchase	-	-	7.4	7.4	7.4	7.4	-	-
7.4	expenditure on selected items by place of purchase		7.4	-	-	-	-	-	-
..	expenditure on petrol, diesel and other motor oils by place of purchase	-	-	7.5	7.5	7.5	7.5	-	-
..	selected household goods and personal goods and services by place of purchase	-	-	7.6	7.6	7.6	7.6	-	-
..	selected regular purchases by place of purchase	-	-	7.7	7.7	7.7	7.7	-	-
7.5	expenditure on clothing and footwear by place of purchase	7.5	7.5	7.8	7.8	7.8	7.8	-	-
8	**Household income**								
8.1	by household composition	8.1	8.1	8.1	8.1	8.1	8.1	8.1	8.1
8.2	by age of household reference person	8.2	8.10	-	-	-	-	-	-
..	by age of head of household	-	8.2	8.2	8.2	8.2	8.2	8.2	8.2
8.3	by income group	8.3	8.3	8.3	8.3	8.3	8.3	8.3	8.3
8.4	by household tenure	8.4	8.4	8.4	8.4	8.4	8.4	8.4	8.4
..	by economic status of head of household	-	8.5	8.5	8.5	8.5	8.5	8.5	8.5
..	by occupational grouping of head of household	-	8.6	8.6	8.6	8.6	8.6	8.6	8.6
8.5	by Region	8.5	8.7	8.7	8.7	8.7	8.7	8.7	8.7
8.6	by GB urban/rural areas	8.6	8.8	-	-	-	-	-	-
8.7	by socio-economic class	-	-	-	-	-	-	-	-
8.8	1970 to 2002-03	8.7	8.9	8.8	8.8	8.8	8.8	8.8	8.8
..	by economic activity status of household reference person	-	8.11	-	-	-	-	-	-
..	by occupation of household reference person	-	8.12	-	-	-	-	-	-

.. Tables do not appear in the 2002-03 report

1 Household reference person replaced head of household in 2001-02
2 Up to 1991 region tables covered two-year periods.
3 Up to 1995-96 region tables related to Standard Statistical Regions, tables from 1996-97 relate to Government Office Regions.

2002-03 tables		Table numbers in reports for							
		2001-02[1]	2000-01	1999-2000	1998-99	1997-98	1996-97	1995-96	1994-95
9	**Households characteristics and ownership of durable goods**								
9.1	households	9.1	9.1	9.1	9.1	9.1	9.1	9.1	9.1
9.2	persons	9.2	9.2	9.2	9.2	9.2	9.2	9.2	9.2
9.3	percentage with durable goods 1970 to 2002-03	9.3	9.3	9.3	9.3	9.3	9.3	9.3	9.3
9.4	percentage with durable goods by income group & hhld composition	9.4	9.4	9.4	9.4	9.4	9.4	9.4	9.4
9.5	percentage with cars	9.5	9.5	9.5	9.5	9.5	9.5	9.5	9.5
9.6	percentage with durable goods by UK Countries and Government Office Regions	9.6	9.6	9.6	9.6	9.6	9.6	9.6	9.6
9.7	percentage by size, composition, age, in each income group	9.7	9.7	9.7	9.7	9.7	9.7	9.7	9.7
..	percentage by occupation, economic activity, tenure in each income group	-	9.8	9.8	9.8	9.7	9.7	9.7	9.7
9.8	percentage by economic activity, tenure and socio-economic class in each income group	9.8	-	-	-	-	-	-	-

.. Tables do not appear in the 2002-03 report
1 Household reference person replaced head of household in 2001-02

Acknowledgements

Editor: Anthony Craggs
Production team: Kay Joseland
Tim Samuels
Sandy Lim
Ann Turner
Ramona Insalaco
Palvi Shah
Dina Dimou
David Devore
Karen Irving
Lynne Rhodes
Chris Kirri
Tony King

Field team and interviewers
Coders and Editors

Cover artwork: Shain Bali